Feed Management in Intensive Aquaculture

Feed Management in Intensive Aquaculture

Stephen Goddard

Fisheries and Marine Institute
Memorial University
Newfoundland, Canada

CHAPMAN & HALL

THOMSON PUBLISHING

New York · Albany · Bonn · Boston · Cincinnati · Detroit
London · Madrid · Melbourne · Mexico City · Pacific Grove
Paris · San Francisco · Singapore · Tokyo · Toronto · Washingto

Cover design: Curtis Tow Graphics

Copyright © 1996 by Chapman & Hall

Printed in the United States of America

Chapman & Hall
115 Fifth Avenue
New York, NY 10003

Chapman & Hall
2-6 Boundary Row
London SE1 8HN
England

Thomas Nelson Australia
102 Dodds Street
South Melbourne, 3205
Victoria, Australia

Chapman & Hall GmbH
Postfach 100 263
D-69442 Weinheim
Germany

Nelson Canada
1120 Birchmount Road
Scarborough, Ontario
Canada M1K 5G4

International Thomson Publishing Asia
221 Henderson Road #05-10
Henderson Building
Singapore 0315

International Thomson Editores
Campos Eliseos 385, Piso 7
Col. Polanco
11560 Mexico D.F. Mexico

International Thomson Publishing–Japan
Hirakawacho-cho Kyowa Building, 3F
1-2-1 Hirakawacho-cho
Chiyoda-ku, 102 Tokyo
Japan

1 2 3 4 5 6 7 8 9 10 XXX 01 00 99 98 97 96

Library of Congress Cataloging-in-Publication Data

Goddard, Stephen.
    Feed management in intensive aquaculture / Stephen Goddard.
        p.   cm.
    Includes bibliographical references and index.
    ISBN 0-412-07081-2 (alk. paper)
        1. Fishes--Feeding and feeds.   2. Shrimps--Feeding and feeds.
    3. Fish-culture.   4. Shrimp culture.   I. Title.
    SH156.G63   1996
    639.3--dc20   96-34801
                                                                                CIP

To order this or any other Chapman & Hall book, please contact **International Thomson Publishing, 7625 Empire Drive, Florence, KY 41042.** Phone: (606) 525-6600 or 1-800-842-3636. Fax: (606) 525-7778. e-mail: order@chaphall.com.

For a complete listing of Chapman & Hall titles, send your request to **Chapman & Hall, Dept. BC, 115 Fifth Avenue, New York, NY 10003.**

Contents

\\\\\\\

CONTENTS

CONTENTS

CONTENTS

Preface

\\\\\\

This book has been written as a guide to the management and use of formulated feeds in intensive fish and shrimp culture. While its focus is on the use of commercially produced feeds in intensive production systems, it is anticipated that many of the practical issues covered will be of equal interest to those fish farmers who make their own feeds and to those who use formulated feeds in less intensive systems. Feeds and feeding are the major variable operating costs in intensive aquaculture and the book is primarily intended to aid decision making by fish farm managers in areas of feeding policy.

The dramatic increases in aquaculture production seen over the past 15 years have been made possible, in large part, by gains in our understanding of the food and feeding requirements of key fish and shrimp species. A global aquaculture feeds industry has developed and a wide range of specialist feeds is now sold. The new options in feeds and feeding systems, which are becoming available, necessitate continual review by farmers of their feeding policies, where choices must be made as to appropriate feed types and feeding methods. While growth rates and feed conversion values are the prime factors of interest to farmers, other important issues, such as product quality and environmental impacts of farm effluents, are also directly related to feed management practices. While there is an extensive literature dealing with the known nutritional requirements of farmed fish and shrimp, less attention has been focused on feed management practices. However, as the various sectors of the industry have matured, and markets have become more competitive, it has become evident that farmers seeking to control their production costs, in order to remain competitive, must critically examine the selection, use, and performance of aquaculture feeds.

While this book emphasizes the practical issues of feed management, it also seeks to provide insight into the biological and environmental factors that underlie the feeding responses of fish and shrimp, and which must be taken into account when determining appropriate feeding policies for specific farming operations. No attempt has been made here to describe in detail the feeding regimens used for any particular species. The examples used throughout the text have been chosen from many different sectors of the industry in an attempt to illustrate the general principles of feed management.

The names of fish and shrimp in common usage within the industry are used throughout the text. Scientific names are included where species are first named, or where there may be some ambiguity concerning the common name of a particular species. The term shrimp is used to refer to members of the marine Penaeidae (e.g., the giant tiger shrimp, *Penaeus monodon*), and prawn to members of the freshwater Palaemonidae (e.g., the giant freshwater prawn, *Macrobrachium rosenbergii*).

Stephen Goddard

Acknowledgments

\\\\\\\

I should like to acknowledge the assistance of the many fish and shrimp farmers who have generously given of their time to describe to me the feeding operations on their farms. It is from these observations and discussions that the idea for this book first developed, and upon which the selection of much of the material has been based.

I also acknowledge the valuable support and encouragement of my colleagues and graduate students at Memorial University. I particularly thank the Office for International Programmes for its continued support of my work overseas. I also thank Wendy Thistle and Karen Collins of the School of Fisheries for all their assistance in the preparation of figures and organization of the manuscript.

Finally I thank Maria Teresa, Benjamin and Clare, for all their understanding and patience, without which this book could not have been written.

Feed Management in Intensive Aquaculture

1

`\ \ \ \ \ \`

Feeds in Intensive Aquaculture

INTRODUCTION

The profitable use of formulated aquaculture feeds in intensive aquaculture is well established. Increased understanding of the nutritional requirements of certain fish and shrimp species, coupled with improvements in feed manufacturing technology and feeding techniques, have been central to the expansion of modern aquaculture. While our knowledge of the nutritional requirements for most farmed species remains far from complete, a global aquafeeds industry has developed and formulated feeds are commercially produced for a diverse range of species (Table 1-1). In addition, many farmers prepare their own feeds from locally available ingredients, using formulations developed and circulated by development agencies (New, 1987; Tacon, 1987, 1988; New, Tacon, and Csavas, 1993).

Formulated feeds are used either to supplement or to replace natural feeds in the diets of farmed fish and shrimp. An overall trend in global aquaculture is toward the increased intensification of culture methods. This trend is in large measure a direct result of the widespread availability of aquaculture feeds, paralleled by improved methods of water quality management, disease control, and general techniques in husbandry. The development of culture techniques for some species has been based from the onset on the use of feeds to support intensive culture, for example, the farming of Atlantic salmon (Fig. 1.1). Elsewhere feeds have been introduced into traditional methods of fish and shrimp culture to support more intensive culture techniques with higher production densities. This trend can be seen, for example, in giant tiger shrimp culture in Southeast Asia, and in the culture of the tilapias throughout the tropics and subtropics.

1

Table 1-1   Examples of Species of Fish and Shrimp Grown in Semi-intensive and Intensive Systems for Which Formulated Feeds are Produced Commercially

Cold-water Finfish		Warm-water Finfish	
Rainbow Trout	*Oncorhynchus mykiss*	Asian seabass	*Lates calcarifer*
Atlantic Salmon	*Salmo salar*	Milkfish	*Chanos chanos*
Pacific Salmon	*Oncorhynchus sp.*	Tilapia	*Oreochromis sp.*
Arctic charr	*Salvelinus alpinus*	Channel catfish	*Ictalurus punctatus*
European eel	*Anguilla anguilla*	African catfish	*Clarius gariepinus*
Sea bass	*Dicentrarchus labrax*	Common carp	*Cyprinus carpio*
Gilthead bream	*Sparus aurata*	Japanese eel	*Anguilla japonica*
European turbot	*Scophthalmus maximus*		
Atlantic cod	*Gadus morhua*	**Warm-water Shrimp**	
Striped bass	*Morone saxatilis*		
Ayu	*Plecoglossus altivelis*	Kuruma shrimp	*Penaeus japonicus*
		Giant tiger shrimp	*Penaeus monodon*
		Chinese white shrimp	*Penaeus chinensis*
		Pacific white shrimp	*Penaeus vannamei*
		Warm-water Prawn	
		Giant freshwater prawn	*Macrobrachium rosenbergii*

INTENSIFICATION

Although the terms extensive, semi-intensive, and intensive are commonly used to define culture methods, these are arbitrary terms, and in practice the distinctions are often less than clear. They are, however, generally linked to the level of inputs of feed and/or fertilizer, and to the stocking densities of fish and shrimp that can be supported.

Extensive Culture

The bulk of fish and shrimp produced in aquaculture are grown in extensive and semi-intensive pond farming systems. In extensive pond culture production is based on the use of organic and inorganic fertilizers. Fertilization of ponds promotes the growth of simple plants which form the base of the pond food chain. Fish or shrimp stocked in such ponds feed within the food chain on phytoplankton, zooplankton, bottom-dwelling invertebrates, and smaller fish. Management practices differ considerably in the extent and type of fertilizers that are used. At its most effective, extensive aquaculture is integrated with other types of crop and livestock production (Little and Muir, 1987). In these systems animal manures and agricultural by-products are used as sources of carbon, nitrogen, and essential minerals to stimulate the primary growth of

Fig. 1.1. The use of nutritionally complete formulated food. Refilling a floating automatic feed dispenser at an Atlantic salmon farm in British Columbia, Canada.
Photo courtesy of Andrea Bezerra

phytoplankton in the presence of sunlight. A range of natural food organisms offers the opportunity to culture several different species of fish, which feed on different prey organisms, within the same pond environment. This practice of polyculture considerably increases the production capacity of the ponds. Fish farming of this type is most commonly carried out in freshwater ponds in tropical and subtropical regions where scales of production range from small artisanal units, supporting families and communities, to large commercial ventures.

Semi-intensive Culture

To increase the production of fish from pond systems, beyond the level supported by the availability of natural food, supplementary foods may be used. Depending on the species of fish or shrimp being farmed, supplementary feeds range from cereals and agricultural and fishery by-products to formulated feeds. Formulated supplementary feeds are generally based on regionally available ingredients and are either prepared by the farmer, or purchased from local manufacturers. Supplementary feeds are often nutritionally incomplete and would be inadequate as a sole source of food. Their function is to provide additional major nutrients and to complement the essential nutrients which the fish or shrimp obtain by consuming natural food organisms. Where cost factors permit, more expensive, nutritionally complete feeds are also used in semi-intensive culture to supplement natural food production.

Widespread culture of the freshwater carps—silver carp, *Hypophthalmichthys molitrix*; common carp, *Cyprinus carpio*; grass carp, *Ctenopharyngodon idella*; and bighead carp, *Hypophthalmichthys nobilis*—is conducted either in extensive or semi-intensive pond systems. Production figures for the world's major cultured species are shown in Figure 1.2. Silver carp production exceeds that of any other finfish with an estimated production of almost 1.5 million metric tonnes (MT) in 1991.

Intensive Culture

In more intensive culture systems there is decreased dependence on the availability of natural food and greater or total dependence on the use of nutritionally complete formulated feeds. In intensive aquaculture the densities of fish that can be maintained in pro-

Fig. 1.2. The major cultured fish and shrimp species. Values are given in metric tonnes and as % of total. (From Tacon, 1993).

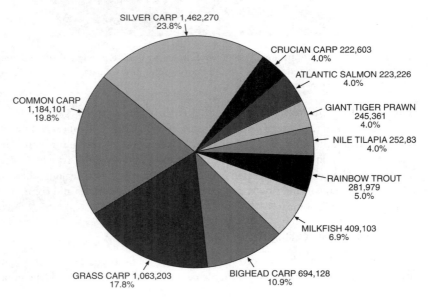

SILVER CARP 1,462,270
23.8%

CRUCIAN CARP 222,603
4.0%

ATLANTIC SALMON 223,226
4.0%

COMMON CARP
1,184,101
19.8%

GIANT TIGER PRAWN
245,361
4.0%

NILE TILAPIA 252,83
4.0%

RAINBOW TROUT
281,979
5.0%

MILKFISH 409,103
6.9%

GRASS CARP 1,063,203
17.8%

BIGHEAD CARP 694,128
10.9%

duction systems are no longer limited by the production of natural food organisms but instead by the tolerance of the species and its ability to grow at higher stocking densities, and for the farmer to maintain adequate water quality. This method of aquaculture is seen in salmon farming in northern Europe, catfish farming in the United States, sea bass and sea bream farming in southern Europe, and eel farming in Japan.

PRODUCTION SYSTEMS

There are numerous examples of production systems to illustrate the points outlined above. Culture methods for the tilapias provide useful illustration since they are farmed in a variety of systems ranging from extensive to highly intensive. The majority of species used in aquaculture belong to the genus Oreochromis (Trewavas, 1983) and include *Oreochromis niloticus, O. aureus, O. hornarum, O. mossambicus* and *Tilapia zilli,* as either pure or hybrid strains (Luquet, 1991) (Fig. 1.3). In extensive culture systems these species feed variously on a range of plankton, benthic invertebrates, vegetation, and algae. These feeding habits were recognized by early fish culturists as an important characteristic of the group and culture methods were initially developed based on the enhancement of natural feed supplies of ponds, using lime and various fertilizers (Fig. 1.4) (Hickling, 1971). In some regions a natural development followed in the use of supplementary feeds in order to increase yields from given pond areas. The use of supplementary feeds is linked to regional economics, since it involves greater cost and

Fig. 1.3. Tilapia hybrids from semi-intensive aquaculture ponds in Ceara, Northeastern Brazil.

Fig. 1.4. A tilapia production pond. A pond used for the semi-intensive culture of tilapia. Lime is being applied to increase the soil pH as an aid to enhancing the productivity of natural food organisms when the pond is refilled.

more labor, which must be reflected in greater yields and income (Fig. 1.5). The use of supplementary feeds is now a common practice in the pond culture of tilapias (Fig. 1.6), and relatively sophisticated systems have been developed to increase yields (De-Silva, 1989). The factors influencing productivity in such systems are complex and require skilled management practices in order to balance the various inputs of fertilizer and feed with optimal fish stocking rates.

Intensive culture of tilapias is less common but accounts for a growing proportion of world production. Intensive culture is carried out in raceway systems (Fig. 1.7) or tanks, with flowing water or in floating cages or enclosures (Fig. 1.8). In these systems the fish have little or no access to natural food organisms and must be given nutritionally complete feeds. Water exchange rates are controlled to maintain adequate levels of dissolved oxygen and to dilute metabolic wastes, while high stocking densities are used to optimize the production capacity of the farm. Table 1-2 shows the range of potential yields from the various systems.

Fig. 1.5. Rice bran. A by-product of rice milling, widely used as a feed supplement in semi-intensive aquaculture.

The levels of intensification adopted in aquaculture relate to the operating objectives of the farm. Intensive systems are generally, although not exclusively, used for the production of high-valued species. This is necessary to justify their relatively high capital and operating costs. Such farms are usually operated to generate revenue through national and international sales of fish and shrimp. In contrast, extensive aquaculture systems are generally established to meet local and regional needs for fish supplies. Semi-intensive farms can fulfill either, or a combination, of these roles. It is common in many regions to find a variety of culture methods employed for the same species and operating side by side.

INTENSIFICATION IN SHRIMP AQUACULTURE

The origins of shrimp aquaculture are seen in the coastal brackish water ponds of Southeast Asia (Fig. 1.9). Some of the earliest records are found in Indonesia, where the operation of tidal ponds (tambaks) can be dated back as early as the fourteenth century. In the traditional operation of tambaks, post-larvae and juveniles enter the pond with the tidal inflow during migratory activities, where they are then retained by mesh gates and screens. The animals feed on the natural algae, plankton, and diverse benthic organisms which proliferate in the warm shallow ponds, and are later harvested at the en-

Fig. 1.6. Hand-feeding a supplementary feed mix to Nile tilapia in semi-intensive culture ponds at Kisumu, Kenya.

trance as they seek to return to the ocean. Production of populations of mixed shrimp species in these systems is highly variable, and is most frequently a secondary crop to a main harvest of milkfish. Production may reach 200–250 kg/ha/crop, with the possibility of two crops per year under favorable conditions.

The trend to more intensive shrimp aquaculture parallels that seen in finfish culture. Stimulated by research advances in the domestication of the kuruma shrimp in Japan, more intensive culture methods were developed for a range of shrimp species. At present most of the world's farmed shrimp are grown in semi-intensive systems in which they feed on a combination of natural prey organisms and formulated feeds. The giant tiger shrimp *Penaeus monodon* (Fig. 1.10) is the most widely cultured shrimp species accounting for over one third of global farmed shrimp. Production levels are variable but are typically in the range 1.5–2.0 tonnes/ha/year. Intensive systems are also used in shrimp aquaculture where high density production is supported by greater use of formulated feeds. These systems may yield up to 10 tonnes/ha/year, based on two production cycles (Fig. 1.11)

The intensification of shrimp production methods has raised problems in some regions, with dramatic production losses, resulting from pollution and the spread of diseases. Poor feeding practices can be a major problem in intensive shrimp culture, creating pollution both within ponds and in neighbouring coastal waters, and promoting an environment in which shrimp are stressed and where opportunistic pathogens thrive.

Fig. 1.7. Intensive production of tilapia. Raceway systems for the production of fingerlings in Java, Indonesia.

Intensive shrimp culture in Taiwan suffered near-collapse in 1988 and 1989, when farmers were confronted with poor growth and high mortality rates. The onset of disease and problems of water supply caused production in that country to fall from 100,000 tonnes in 1987 to 30,000 tonnes in 1988 and to 20,000 tonnes in 1989. Despite these, and similar problems elsewhere, intensive culture systems are being widely introduced, and contribute an increasing proportion of the world's farmed shrimp. In Thailand, one of the leading producers of farmed shrimp, over 90% of new farms built between 1989–91 were planned for intensive operations (Csavas, 1994).

Distinctions between semi-intensive and intensive production methods are less marked in shrimp farming than in fish farming, since natural food organisms play a significant role in both systems, but the overall feeding dynamics are poorly understood. Experiments involving the use of stable carbon isotopes have shown that 60–70% of shrimp *(Penaeus vannamei)* growth in semi-intensive systems resulted from the consumption of natural food organisms, while formulated feeds accounted for the remain-

Fig. 1.8.   Intensive production of tilapia. Floating cages in a freshwater lake in Java, Indonesia.

ing 30–40% (Anderson et al., 1987). Since there is no ready measure of the contribution of natural foods in the diets of farmed shrimp, calculations by the farmer of conversion rates based on the quantities of formulated feeds used, may overestimate the efficiency of feed use. More research is needed in order to determine both the uptake levels of formulated feeds, and the fate of uneaten food within the pond ecosystem.

FEED DEVELOPMENT

The requirements of intensive aquaculture for nutritionally complete feeds have stimulated considerable research and development activity in the fields of fish nutrition and feed technology. During the late 1960s and the 1970s intensive aquaculture of a number of species was developing. Rainbow trout and Atlantic salmon farming in Europe, channel catfish farming in the United States and kuruma shrimp farming in Japan are key examples, since the subsequent growth of these industries was dependent on the

Table 1-2 Examples of Tilapia Yields from Various Production Systems[1]

	Extensive		Yield
Production Costs	No Inputs	→	150 kg/ha/250 days
	Organic Fertilizer		1000 kg/ha/250 days
	Semi-intensive		
Water Requirements	Rice Bran	→	1650 kg/ha/180days
	Supplementary feed		3500 kg/ha/180days
	Intensive		
Management Input	Complete feeds: raceways/tanks	→	$70–100 + kg/m^3/yr$ [2]
	Complete feeds: cages		$50–100 kg/m^3/yr$ [2]

[1]Details collected from various sources in Nigeria, Brazil, Kenya, Indonesia, and USA.
[2]Yields of fish from intensive systems are generally expressed in terms of production/water volume/time. Yields from less intensive systems are expressed in terms of production/pond surface area/time.

availability of nutritionally complete feeds. The use of formulated feeds is now extending to a steadily growing range of farmed species.

The development of formulated feeds for trout and salmon has been at the forefront of the growth of the aquaculture feeds industry. Our understanding of the nutritional requirements of these species exceeds that available for most other species, and manufacturers have produced an increasingly sophisticated range of trout and salmon diets. Prior to the 1950s trout and salmon grown in hatcheries in Europe and the United States, for restocking programs, were fed a range of wet feeds prepared from minced trash fish and animal offal. Freshwater farming of rainbow trout in the Jutland region of Denmark, was one of the first intensive industries to be developed. This pioneer industry, first introduced at the turn of the century, grew rapidly during the 1950s and was initially based on the use of by-catch marine fish for food. The proximity of the farms to the Danish North Sea fishing ports made daily delivery of fresh feeds to the farms possible.

A significant step in the development of formulated feeds came with the introduction of the Oregon Moist Pellet in the early 1950s. This feed, based on a mixture of wet and dry ingredients, was developed by researchers at Oregon State University and the Oregon Fish Commission. The wet fish products used in this feed were pasteurized, reducing the risk of disease transmission, and vitamin and mineral supplements were included. The Oregon Moist Pellet is still widely used in the production of Pacific salmon smolts in freshwater hatcheries. It is manufactured and delivered frozen to the hatcheries and then thawed prior to use. Parallel with the development of moist feeds for use at salmon hatcheries, researchers in Europe and the United States were developing dry feed formulations for trout. Work at the trout and salmon hatchery at Cortland, New York, under the supervision of A.V. Tunison and A.M. Phillips, pioneered this field (Rumsey, 1994), and dry feeds became commercially available in the late 1950s. During the 1960s as rainbow trout farming, followed by Atlantic salmon farming, devel-

Fig. 1.9 Shrimp farms. Modern intensive and semi-intensive farms set among traditional tambaks in Java's coastal zone.

Fig. 1.10 Farmed shrimp. Juvenile *Penaeus monodon* from semi-intensive culture ponds in Indonesia.

oped in Europe, wet, moist and dry feeds were all used. The choices of farmers were based on relative costs and regional availability.

In the southeastern United States the rapid growth of intensive channel catfish farming paralleled trout and salmon farming in Europe. The detailed nutritional requirements of catfish were gradually identified and the formulations of dry feeds were gradually modified, in the light of new information, to the present stage where nutritionally complete, dry feeds are produced economically for use in intensive culture. In Japan the development of intensive culture methods for the kuruma shrimp, *Penaeus japonicus*, was similarly supported by basic nutrition research. In the early development stages during the 1960s cultured shrimp were fed on fresh food, comprising bivalves, squid or scrap fish. These practices were followed by the gradual introduction of formulated feeds during the 1970s as more details of the nutritional requirements of shrimp became available.

Research into the nutritional needs of a growing number of fish and shrimp species has followed these early developments. While the nutritional principles to be followed in feed development are similar for all animals, significant differences have been demonstrated between fish and between shrimp species. Some 40 essential dietary nutrients, including amino acids, fatty acids, vitamins and minerals, are required by all animals. The quantitative requirements differ between species, however, and this presents

Fig. 1.11   An intensive shrimp farm. Production methods are dependent on the use of formulated feeds and water quality management.

continuing challenge to fish nutritionists and feed manufacturers as new species are investigated. At present more than 100 fish and shrimp species are cultured worldwide.

Most intensively farmed species are carnivorous, with high requirements for animal protein and energy in their diets. This has necessitated the development of nutrient-rich feeds in some sectors of intensive aquaculture, and dependence on the use of expensive ingredients of high nutrient quality, such as fish meals and marine oils. Feed formulations for some fish species have changed significantly in recent years. These changes have resulted from greater understanding of the fish's nutrient requirements, improved quality of ingredients and manufacturing methods, and responses to environmental concerns and pollution legislation. Table 1-3 outlines examples of the major changes which have occurred in the development of formulated feeds for rainbow trout since their original manufacture. These changes show significant reduction in the inclusion of phosphorus and the increased use of lipids to supply energy.

Feed development for aquaculture species is complicated by the need to produce diets which are water stable. This requirement is particularly critical for shrimp and those fish species which feed slowly, over a period of several hours. Diets must be effectively bound to prevent breakup in the water and leaching of certain nutrients must be minimized. Water soluble vitamins, free amino acids and some minerals are rapidly lost from aquaculture feeds once the feeds are immersed in water. These are problems peculiar to aquaculture feeds and have necessitated a reexamination of ingredients, feed formulations and manufacturing techniques. Some conventional livestock feed ingre-

Table 1-3   Changes in the Composition of Dry Trout Feeds in Europe, 1950–90

	Protein %	Oil %	Carbohydrates %	Phosphorus %	Energy Kcal
Wet feed					
1950	55	26	1.6	1.6	4300
Dry feed					
1950–59	35	5	30	2.5	2230
1960–69	40	7	23	2.0	2600
1970–79	53	11	12	1.5	3100
1980–89	56	20	10	1.1	3800
1989–90	42	24	19	0.9	4300

From Jensen, 1991.

dients have been replaced in some aquaculture feeds. For example, protected forms of vitamin C, with reduced solubility, are now widely used in aquaculture feeds. A number of other physical and chemical characteristics distinguish aquaculture feeds from conventional livestock feeds. It is possible, through certain methods of feed manufacture, to produce pellets of varying density. Hence fish farmers can select pellets which float or sink at different rates, and are appropriate for use with particular species in particular types of rearing systems. Shrimp farmers use relatively fast sinking feeds since shrimp are bottom feeders, while in contrast, many catfish farmers opt to use floating pellets in order that they can observe the feeding response of the catfish as they come to the surface for food.

The high nutrient specifications seen in aquaculture feeds, and the use of specialized manufacturing techniques are reflected in the high costs of most aquaculture feeds when compared with the costs of conventional livestock feeds. These costs are further inflated by the needs of feed manufacturers to overcompensate levels of certain essential ingredients, either because absolute requirements are not fully known, or where undetermined amounts of soluble ingredients may be lost from pellets before they are eaten.

FEED INGREDIENTS

Fish meals are major components of aquaculture feeds. In most aquaculture feeds they are the main source of protein and may constitute up to 60% of the total diet (Fig. 1.12). With few exceptions, intensive aquaculture has become heavily dependent on the use of high grade fish meals. Channel catfish farming in the United States is one notable exception. Progressive research into the nutritional requirements for this species has resulted in the use of formulated feeds containing less than 10% fish meal, but which contain soybean meal and grains as the major ingredients.

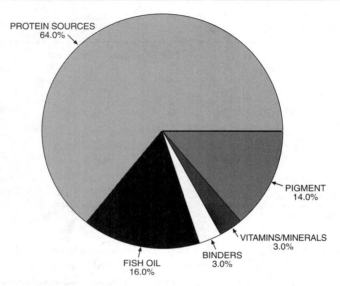

Fig. 1.12 Feed cost breakdown. An example of the percentage cost breakdown of a commercial, extruded Atlantic salmon grower diet employed in British Columbia, Canada. (From Prendergast et al., 1994)

Fig. 1.13 Fish meal consumption. The use of fish meal by various sectors of the aquaculture industry in 1992. Values are given as % and '000 metric tonnes. (From Tacon, 1993)

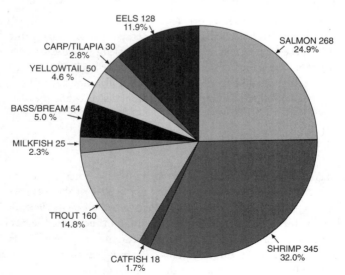

In 1992 some 6 million tonnes of fish meal were produced, of which 1 million tonnes were used in aquaculture (Fig. 1.13). This represented 16% of the total global production of fish meal; a figure predicted to increase up to 20–25% of the total by the year 2000 (Rumsey, 1993). Of the various grades of fish meal available, those used in aquaculture feeds are typically manufactured from whole fish. Meals manufactured from fish offal and processing waste, while cheaper, often lack the nutritional profile suitable for use in formulated fish and shrimp feeds.

In the manufacturing of fish meal, oil is separated then the fish is dried. Various methods are used for drying fish, using either hot air or steam. Low temperature (LT) fish meals are increasingly used in aquaculture feeds since there is evidence that these meals are of superior quality, in terms of their nutrient availability for fish. LT meals are dried at temperatures below 80°C in comparison with drying temperatures of more than 100°C used in conventional fish meal manufacture. However, the main factor controlling the quality of fish meals is the freshness of the raw material. Fish meals are typically manufactured from oily pelagic fish species, such as herring, caplin, and anchovy. Spoilage of these fish is rapid and production of high specification meals for use in aquaculture entails careful control of the raw products. This generally involves chilling of the fish immediately after capture at sea and reducing storage time to a minimum. Production of high quality fish meals in the tropics poses particular problems with regard to competing demands for the raw materials, spoilage, and subsequent quality control. This has broadened the international trade in fish meals where rapidly emerging feed manufacturing sectors (e.g., domestic shrimp feed manufacturers in Thailand and Indonesia) import high-grade European and South American fish meals in order to meet the nutrient specifications of modern shrimp feeds which cannot readily be met by the use of domestic fish meals.

The international demands for fish meal may represent a future limitation in the growth of intensive aquaculture for carnivorous species. The major producer countries of good quality fish meal—Peru, Chile, Denmark, Japan, Iceland, Norway, USA, and South Africa—are already exploiting pelagic fish stocks to their sustainable limits. Unless new sources can be identified, or use made of the substantial discards which are currently wasted in the fishery, demand may well exceed supply before the end of the decade as shrimp and fish farmers compete with the long-established poultry and livestock industries for a limited resource. Improvements in fish harvesting and processing technologies, coupled with changing consumer demand, may also have some impact on future supplies if some fish species, traditionally used for reduction to fish meal, are used in products for direct consumption.

Alternative Feed Ingredients

If the predicted growth of intensive and semi-intensive aquaculture is to be sustained into the next century more efficient use of resources and a reevaluation of the use of feeds and feed ingredients will be necessary. Alternative ingredients for fish meals are being sought, and may offer a partial solution in the future. The nutritional demands of carnivorous species limit the extent to which plant proteins can contribute to diets formulated for most species of intensively farmed fish and shrimp. Some protein sources from crop plants, such as soybean meal, are, however, used to varying levels in diets for both carnivorous and omnivorous species. Where existing biotechnologies are used to remove enzyme inhibitors and antinutritional factors, soybean, canola (rapeseed) and other cereal proteins can be predicted to play a more significant role in formulated feeds in the future. The use of single cell proteins (SCP), comprised of bacteria or yeasts, grown in industrial-scale fermentation systems, may also increase as replacement ingredients for fish meal in aquaculture feeds.

At present, however, feeds used in the intensive culture of most species contain fish meal as their major ingredient, and their use is expanding rapidly in both intensive and semi-intensive aquaculture. The global production of aquaculture feeds is predicted to increase from its estimated 1990 level of 2.9 million metric tonnes (MT) to 4.6 million MT by the year 2000 (Chamberlain, 1993).

AQUACULTURE FEEDS AND THE ENVIRONMENT

In addition to the questions raised as to the future availability of food resources used in intensive aquaculture, the current usage of aquaculture feeds has also raised environmental concerns. Denmark, one of the first countries to develop intensive aquaculture, was also one of the first to recognize the pollution problems associated with the industry. The close proximity of numerous trout farms, discharging effluent into rivers, created pollution problems which were ultimately addressed by the introduction of controls on the use of feeds by the industry. The use of wet feeds was prohibited and effluent discharge controls were tightened. More recently, legislation has been introduced which sets limits on food conversion values. This latest move puts pressure on the farmers to reduce food wastage by optimizing feed management practices. The experience in Denmark and elsewhere has led to a widespread examination of the chemical and biological nature of effluents from fish farms, and their effects on release into natural water bodies.

The impact of aquaculture on the environment is primarily related to issues of feed management. Overfeeding is the most common source of problems. The composition of feeds and food conversion values affect both the physical and chemical nature of waste materials and the amounts produced. The main waste products associated with feeding are uneaten food, feces, ammonia and carbon dioxide. There may also be antibiotic residues resulting from the use of medicated feeds.

The impact of these materials on the environment is significantly influenced by the water exchange characteristics of the fish or shrimp farm site. Most species thrive in clean, oxygen-rich water, and sites for intensive farms are normally selected on the basis that they have sufficient water exchange to dilute farm effluents to "safe" levels. This is generally the case with metabolites such as ammonia and carbon dioxide. If water is recycled special measures must be taken to remove suspended solids and ammonia, to maintain pH, and to increase dissolved oxygen levels.

Feces and unconsumed food pose special problems in the aquatic environment where they contribute an unnatural loading of organic material into the farm environment that may cause significant changes in water quality. Initial bacterial breakdown of organic materials added to the water in fish farms consumes vital oxygen. Later, if organic materials accumulate to significant levels in the sediment below floating cages, or on the bottom of ponds, the sediment will become anaerobic, releasing toxic hydrogen sulphide and methane gases.

Suspended solid materials in the water column may also have a direct harmful affect on the delicate gill surfaces of fish or shrimp, affecting respiratory function and providing potential sites for bacterial infection. Long-term effects may result from the release of phosphates and nitrates from fish farms. While nontoxic, they can lead to the enrichment or eutrophication of the water. This creates conditions that stimulate the growth of plants and other organisms. These may exert competing oxygen demands with farmed fish stocks and in extreme cases may contribute to the occurrence of potentially harmful plankton blooms.

There are widespread concerns regarding the release of phosphates into the environment. Phosphorus is one of the most important minerals in fish nutrition where it plays vital roles in growth, bone mineralization, and lipid metabolism. Many commercial diets contain phosphorus in excess of the known requirements that have been determined for only a few species. This arises from questions as to the availability of phosphorus from complex ingredients such as fish meal, and limited information as to the essential nature of phosphorus for growing fish. This prompts feed manufacturers to provide excess amounts rather than risk compromising the growth of the fish. This is a common problem in formulating fish feeds and stems from incomplete information as to the finite requirements of the cultured species for which feeds are manufactured. More research is necessary to give more complete understanding of mineral digestion and absorption processes for most farmed species. The introduction of controls on the size of fish farms and on the phosphorus content of farm effluents are measures that have been taken in many countries to reduce its harmful role in the eutrophication of natural waters receiving fish farm effluent.

INDUSTRY GROWTH

World aquaculture production has increased rapidly in recent years, reaching an estimated total of 19.3 million tonnes in 1992 with a value of $32.5 billion (US) (FAO, 1994). These figures include all farmed aquatic organisms, that is, fish, molluscs, crus-

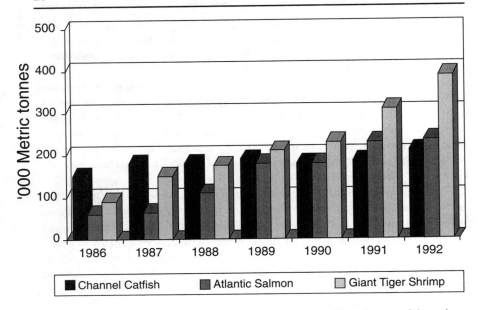

Fig. 1.14   Production figures. The annual increases in total world production of channel catfish, Atlantic salmon, and giant tiger shrimp, 1986–1992. (Data from FAO, 1994)

taceans, and aquatic plants. World capture fishery landings are estimated at 100 million tonnes per year. This figure is predicted to remain relatively stable into the future, hence, any increased future demand for seafood will have to be met by aquaculture.

Production statistics (FAO, 1994) reveal the significant growth of intensive aquaculture during the period 1986–92, and this growth is predicted to continue, albeit at a slower rate (Hempel, 1993). Actual growth over the next decade will be linked to many factors, including global economics, the availability of essential resources, and the success of the industry in addressing environmental concerns. The large-scale commercial development of intensive culture of species such as the channel catfish, Atlantic salmon, and giant tiger shrimp (Fig. 1.14), is already extending to marine finfish species, such as turbot, sea bass, and sea bream, while culture techniques for new aquaculture species, such as Atlantic halibut, African catfish, and Asian sea bass are in the early stages of commercial development. On a broader scale, the use of supplementary feeds in semi-intensive culture continues to expand as farmers seek to increase production from semi-intensive culture of shrimp and prawns, and of such finfish as tilapia, milkfish, and carp.

Management Factors

In developing appropriate feed management policies numerous factors must be taken into account. These include economic, social, biological, and environmental factors,

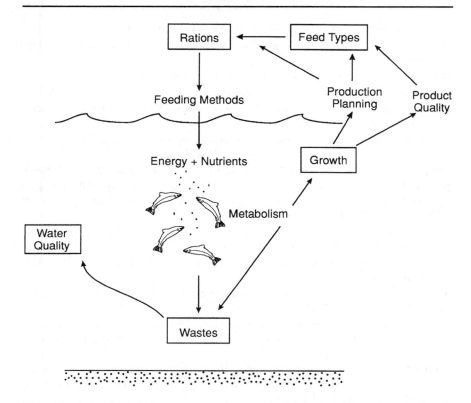

Fig. 1.15   Aquaculture feed management. A summary of the various factors that must be taken into account and their interrelationships.

which in practice, range from feed selection to issues of final product quality. The interrelationships of these factors are illustrated in Figure 1.15 and are discussed in the following chapters.

REFERENCES

ANDERSON, R.K., P.L. PARKER and A. LAWRENCE. 1987. A 13C/12C trace study of the utilisation of presented feed by a commercially important shrimp *Penaeus vannamei* in a pond growout system. Journal of the World Aquaculture Society **18**:148–155.

CHAMBERLAIN, G.W. 1993. Aquaculture trends and feed projections. World Aquaculture **24**:19–29.

CSAVAS, I. 1994. Important factors in the success of shrimp farming. World Aquaculture **25**:34–56.

DESILVA, S.S. 1989. Reducing feed costs in semi-intensive aquaculture systems in the tropics. Naga (ICLARM Newsletter) **12**:6–7.

FAO. 1994. Aquaculture Production. Fisheries Circular. No. 815. Rev. 6.F1D1/C815.

HEMPEL, E. 1993. Constraints and possibilities for developing aquaculture. Aquaculture International 1:2–19.

HICKLING, C.F. 1971. Fish Culture. Faber and Faber, London. p. 317.

JENSEN, P. 1991. Aquafeeds: reducing environmental impact. Feed International August:6–12.

LITTLE, D. and J. MUIR. 1987. A Guide To Integrated Warm Water Aquaculture. Institute of Aquaculture Publications, Stirling.

LUQUET, P. 1991. Tilapia, *Oreochromis spp.* In Handbook of Nutrient Requirements of Finfish, ed. R.C. Wilson, pp.169–180. CRC Press Inc., Boca Raton.

NEW, M.B. 1987. Feed and feeding of fish and shrimp. ADCP/REP/8726:p. 275. FAO, Rome.

NEW, M.B., A.G.J. TACON, and I. CSAVAS (eds.). 1993. Farm Made Aquafeeds. Proceedings of FAO/AADCP Regional Expert Consultation on Farm-Made Aquafeeds, December 14–18, 1992, p. 434. FAO/AADCP, Bangkok.

PRENDERGAST, A.F., D.A. HIGGS, D.M. BEAMES, B.S. DOSANJH, and G. DEACON. 1994. Searching for substitutes–Canola. Northern Aquaculture 10(3):15–20.

RUMSEY, G.L. 1993. Fish meal and alternate sources of protein in fish feeds; update 1993. Fisheries 18:14–19.

RUMSEY, G.L. 1994. History of early diet development in fish culture, 1000 B.C. to A.D. 1955. Progressive Fish-Culturist 56:1–6.

TACON, A.G.J. 1987. The nutrition and feeding of farmed fish and shrimp—A training manual. 2. Nutrient sources and composition. FAO Field Document, Project GCP/RLA/075/ITA, Field Document No. 2, Brasilia, Brazil, p. 129.

TACON, A.G.J. 1988. The nutrition and feeding of farmed fish and shrimp—A training manual. 3. Feeding methods. FAO Field Document, Project GCP/RLA/075/ITA, Field Document No. 7, Brasilia, Brazil, p. 208.

TACON, A.G.J. 1993. Feed ingredients for crustaceans—Natural foods and processed feedstuffs. FAO Fisheries Circular No. 866. FAO, Rome.

TREWAVAS, E. 1983. Tilapiine Fishes of the Genera *Sarotherodon, Oreochromis* and *Danakilia*. British Museum (Natural History), London.

2

`\\\\\\`

Feeding and Diet

INTRODUCTION

An understanding of the natural foods and feeding habits of farmed fish and shrimp can be a significant contributing factor in the planning of effective feeding regimes. While intensive fish farming has demonstrated the ability of some fish species to adapt to a wide range of feeding practices, their natural patterns of feeding behavior, which reflect anatomical and physiological adaptations to their environment and diet, should be taken into account when selecting feeds and planning feeding regimens on farms. In intensive culture systems fish or shrimp are fed on formulated feeds that differ significantly in size, taste or texture from their natural food. While their response to these feeds may be modified by the farm environment it also reflects their natural adaptations in feeding habit.

DIVERSITY

Among the fishes we see the greatest species diversity to be found within the various vertebrate groups. Over 20,000 species have been described, and in the course of their evolution fish have come to occupy a wide range of aquatic habitats extending from tropical fresh waters to cold polar seas. It has been estimated that 41.2% of fish inhabit freshwater, 58.2% are marine, and 0.6% are diadromous, migrating between fresh and saline waters (Cohen, 1970).

Within these diverse habitats fish display a wide range of feeding habits. These are reflected in the structural adaptations of their mouths and teeth, in the functioning of their digestive systems, and in their feeding behavior. The majority of fish species are carnivorous, adapted to feed primarily on animal matter. Relatively few species are omnivores, feeding on a mixed plant and animal diet, or true herbivores, feeding exclusively on plants. Among aquaculture species we can find animals which in nature exploit virtually all types of organic food sources; ranging from fish species which predate on other fish, to detritus feeders which derive their nutrients from decaying plant and animal material (Table 2-1). Although we can broadly categorize aquaculture species as herbivores, omnivores or carnivores, rigid definitions are often inappropriate because some species will feed opportunistically on whatever supply is most abundant. This is particularly true for the fry and juvenile stages of many species where food particle size may be the critical factor in determining food intake. The juvenile stages of most species of fish which are predominantly herbivorous as adults, feed omnivorously on algae and mixed plankton. With few exceptions, fish grown in intensive aquaculture are carnivorous, while most fish grown in extensive and semi-intensive systems are omnivorous and herbivorous species, grown in warm-water systems (Figs. 2.1 and 2.2).

There is considerably less diversity in the feeding habits of those decapod crustaceans of commercial importance in aquaculture. They have been observed to feed opportunistically, in accordance with the relative abundance of prey organisms. As a group the penaeid shrimps are typically carnivorous, feeding on small animals, while freshwater caridean prawns and crayfish feed more omnivorously on a variety of plant and animal material (New, 1995; Goddard, 1988).

The protein and energy requirements of fish irrespective of their natural diet are re-

Table 2-1   The Feeding Habits of Fish Species

Feeding habit	Estimate % of known species[1]	Examples of farmed species
Carnivores	85	Japanese eel, sea bass, sea bream, rainbow trout, Atlantic salmon, yellowtail
Herbivores	6	Grass carp, silver carp, some *tilapias*, milkfish
Omnivores	4	Common carp, channel catfish, some *tilapias*
Detritivores	3	Mud carp (*Cirrhina molitorella*), mullet (*Mugil spp.*)
Scavengers and parasites	2	

[1] Data from Pandian and Vivekanandan, 1985

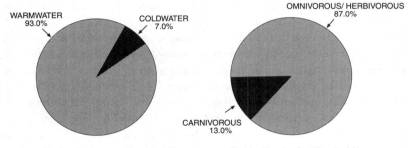

Fig. 2.1 Temperature and feeding habits of farmed fish. Data from 1991 global production statistics. (From Tacon, 1993)

markably similar. Fish which naturally feed on plant or detrital material consume greater volumes of food and, as a result, achieve comparable nutrient intake levels. Fish also show remarkable adaptability in their capacity to ingest and process formulated feeds. Omnivorous species of tilapia and herbivorous milkfish will feed readily on nutrient-dense formulated feeds. The structure and function of their digestive systems influence their patterns of food intake and digestive efficiency, however, and meal sizes and feeding frequencies must be set accordingly.

FEEDING BEHAVIOR

Critical factors in the effective use of feeds are first, the need to ensure that food given to fish and shrimp is consumed as completely and as rapidly as possible, and second, that the food contains the necessary ingredients to meet the metabolic needs of the animal and to support growth. Food not eaten becomes a potential pollutant, while delay in consumption increases the risk that soluble nutrients may leach from the food.

In situations where food wastage results from overfeeding, an examination of feeding methods generally reveals that food is being given to fish or shrimp in excess of levels needed to match their appetite. An understanding of appetite and feeding behavior is therefore of prime importance to fish farmers in determining optimal feeding regimens. There is a need to ensure that feed regimens, meal sizes, feeding frequency, and duration of feeding bouts, are adjusted in order to optimize growth and feed conversion values.

Appetite

A complex of mechanisms, metabolic, neurophysiological, and hormonal, are involved in the regulation of appetite and feeding behavior in fish. Environmental and dietary factors influence appetite, and some routine operations on fish farms, such as grading and sorting fish, may stress fish, temporarily depressing their feeding response and lowering their levels of food intake.

Appetite in fish is taken by most fish farmers as the key indicator by which to judge feeding requirements. For those fish which can be readily observed feeding at the water surface, for example, eels, rainbow trout, and tilapia, an experienced fish farmer can recognize the various stages of the feeding cycle. Initial feeding behavior is characterized by active swimming and searching, followed by vigorous feeding. This phase is followed by a progressive reduction of the feeding response until feeding ceases, and the fish are satiated. After feeding, the fish generally reschool, swimming below the surface.

The physiological processes regulating appetite in fish appear to be similar to those found in the higher vertebrates. At least two types of regulatory feedback mechanisms are believed to operate, a short-term response to stomach distension and a longer-term response to levels of certain nutrients circulating in the blood system. During feeding bouts the brain, linked to receptors in the wall of the stomach or foregut, monitors the degree of distension or fullness as feeding progresses. When the stomach is full, or partially full, satiation centers in the hypothalamic region of the brain respond to the information received, and feeding activity is reduced and eventually stopped. Resumption of feeding activity will not normally reoccur until the stomach has emptied a large proportion of its contents into the intestine. At this time, feeding centers in the hypothalamus, which monitor information from a variety of sources, will trigger appetitive behavior and the search for food. It is also generally accepted (Holmgren et al., 1983; Smith, 1989) that changes in the levels of nutrients—glucose, fatty acids, glycerol, and essential amino acids—absorbed from the digestive tract and circulating in the bloodstream of the fish are monitored by receptors in the brain and in the liver. The physiological mechanisms regulating food intake in fish are poorly understood although various models have been proposed (Vahl, 1979; Grove, 1986).

Measurements of voluntary food intake levels to quantify appetite, have shown that the return of appetite in some carnivorous species, for example, rainbow trout (Grove et al., 1978), flounder, *Limanda limanda* (Gwyther and Grove, 1981) and the European eel (Seymour, 1989), is closely correlated to stomach emptying times. It is still not clear, however, to what extent appetite is controlled by stomach distension in fish, but satiation time and the subsequent return of appetite seem to be the main factors limiting the number of meals a fish will eat each day (Boujard and Leatherland, 1992).

It has been shown that feeding behavior is regulated in accordance with the metabolic needs of the fish. Several studies have shown that fish regulate their food intake to meet their energy requirements (Grove et al., 1978; Jobling, 1980). Assuming the continuous availability of food, fish compensate for low energy density in their food by eating more. The limits of such compensation are set by the physical capacity of the gut. Up to this limit, fish offered diets of different energy density will grow similarly, assuming that their diet contains an adequate balance of nutrients.

Attractants

Feed intake in fish and shrimp is a selective process. Feeds may be readily ingested or rejected, based on the nature of their chemical scents and taste. Even in those species that detect food visually, taste is used in making the final decision as to whether to swallow or reject a particular type of food. The inclusion of chemical attractants in feeds formulated for both fish and shrimp is now widespread. A number of commercial products are available which are variously claimed to increase food intake, with resultant gains in growth rate.

Progress has been made in recent years in identifying those specific chemicals which act as feeding stimulants for fishes. This research has been of particular importance in the development of larval and starter feeds, used to wean larvae from live prey diets onto formulated diets. Feeding stimulants identified for both fish and shrimp possess some general characteristics (Carr, 1982):

- They contain nitrogen.
- They have a low molecular weight (<1000).
- They are nonvolatile and water-soluble.
- They possess both acid and base properties simultaneously (= amphoteric).
- They are widely distributed in plant and animal tissues.

In general carnivorous species show the greatest positive response to alkaline and neutral substances, such as glycine, proline, taurine, valine, and betaine, while herbivores respond more to acidic compounds, such as aspartic and glutamic acids. Of these compounds betaine is the most widely used attractant, in both fish and shrimp feeds used either alone, or in combination with other nitrogenous compounds, such as amino acids and nucleotides.

DIGESTION

Digestion in Fish

Digestive processes in fish are similar to those observed in other vertebrates (Fänge and Grove, 1979; Smith, 1989). The function of the digestive tract is to reduce ingested food, through a process of physical and chemical breakdown, to its constituent molecules which are then absorbed in solution through the gut wall into the blood circulatory system. The digestive tube comprises an inner mucosal layer which contains digestive enzyme and mucus secreting cells, a supportive submucosal layer, smooth muscle layers, which through contractions mix and transport food along the digestive tract, and an outer serosal layer, composed of fibrous connective tissue.

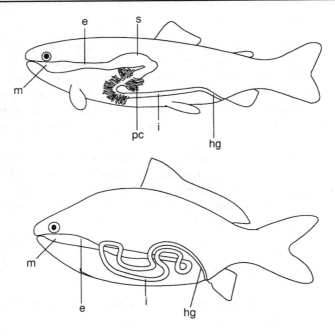

Fig. 2.2 Diagrammatic representations of the alimentary systems of i. a carnivorous fish—rainbow trout; ii. an omnivorous fish—common carp.
e: esophagus; s: stomach; pc: pyloric caecae; i: intestine; hg: hind gut

In adaptation to diet greatest variations are seen in the structure of the mouth and teeth, gill rakers, the pharynx, the stomach (if present), and in the length of the intestine. Fish digestive systems, representative of carnivorous and omnivorous feeding habits, are illustrated in Figure 2.2.

Carnivorous fish typically posses a large, muscular walled stomach and a relatively short intestine. In predatory fish this structure correlates with a feeding habit based on the consumption of relatively large items of prey, followed by a gradual process of digestion. The outer layer of the food mass lying in the stomach is gradually eroded and food is continually passed into the intestine for further digestion and absorption. Active feeding periods are spaced at intervals throughout the day, with the stomach acting as a temporary storage organ, maintaining a flow of nutrients into the intestine. In contrast, omnivorous and herbivorous species feed on materials with a lower nutrient density than do carnivores. For some species this necessitates almost continuous feeding throughout the day. These fish typically posses a long intestine and their stomachs, if present, may be poorly developed. For herbivorous fish, feeding on plant material, only part of the ingested food is retained in the stomach while the remainder is passed rapidly into the intestine. Pharyngeal teeth are present in some species, such as the common carp and some tilapias. These function to compress and reduce food prior to chemical digestion in the intestine. Other species, such as milkfish, have a gizzard-like structure in the foregut, which functions to reduce food particle size before enzymic digestion in the long intestine.

Digestion of protein and lipids in carnivorous fish is efficient, resulting in high levels of nutrient absorption. Their digestive systems, however, are unable to process or absorb high levels of dietary carbohydrate. They secrete insufficient enzymes to break down complex carbohydrates, such as starch, and their digestive tract is too short, and gut clearance times too rapid, to accommodate prolonged retention time for more efficient digestion and absorption. Starch pretreated through extrusion processes in feed manufacture is more digestible than raw untreated starch, and is a common ingredient in feeds formulated for most fish and shrimp species, where it functions as both a binding agent and as a source of energy.

The presence of large amounts of indigestible materials, such as plant fibers, in the natural diets of omnivorous and herbivorous species, is reflected in lower levels of digestibility and absorption, per unit of food ingested, than is typical in carnivores. These fish can, however, digest and absorb certain carbohydrates more efficiently than carnivorous species. The efficiency of digestion in fish is not significantly affected by environmental factors such as temperature or salinity, or by biotic factors such as age or size.

Feeding and Digestion in Shrimp

The various species of penaeid shrimp that are grown on farms are typically carnivorous in feeding habit. They feed primarily on small animals, such as molluscs, crustaceans, and polychaetes which live in, or on, the surface layer of the pond bottom. They also consume some plant material, particularly in the juvenile stages, and plant detritus, from which nutrients are probably derived from the associated micro-organisms. Particular food items may predominate in the diets of cultured shrimp. Examination of the food, feeding rhythmicity, and feeding behavior of the southern brown shrimp, *Penaeus subtilis*, grown in semi-intensive culture ponds in northern Brazil revealed a preference of this species for benthic polychaetes. Analysis of stomach contents and stable carbon isotope ratios, revealed that polychaetes constituted 32.5% of total food intake, while formulated food accounted for 15.6%. Of particular interest from this study, sampling of stomach contents over 24 h periods, showed that application of formulated feeds to the ponds promoted the consumption of natural prey organisms, presumably by stimulating foraging activity (Nunes, Goddard and Gesteira, unpublished observations). Careful monitoring of the populations of prey organisms is a vital component of feed management when supplementary feeds are used in semi-intensive culture. Fluctuations in the abundance of prey organisms should be accounted for in the calculation of formulated feed ration size throughout the rearing cycle.

While often described as scavengers, observations of shrimp feeding in aquaria have revealed that some species display a marked preference for fresh food, over food that has been soaked in water for several hours (Hill and Wassenberg, 1987). These observations probably reflect the response of the shrimp to chemical stimuli. Low concentrations (10^{-5}–10^{-6}M) of amino acids have been shown to elicit a feeding response in *Penaeus merguiensis*, which resulted in the shrimp walking faster, probing deeper into the substrate, and spreading their limbs wider (Hindley and Alexander, 1978). At exposure to higher concentrations of amino acids (10^{-1}–10^{-2}M), shrimp display masticatory movements of their mouthparts. These observations have led to the suggestion that shrimp have two levels of sensitivity to food stimuli: a sensitivity to low concentrations, which allows the shrimp to detect food at a distance, and a sensitivity to high concentrations, which enables the shrimp to detect food from other materials prior to ingestion (Hindley, 1975a,b).

Prey is detected by sensory receptors on the first chelae. If capture is successful the food organism is grasped by the chelae and transferred to the mouthparts. The maxillae and first and second pairs of maxillipeds are adapted to enmesh and hold the prey prior to ingestion. Large items of food are crushed or cut by the mandibles, while non-food items are rejected. Secretions of mucus, from glands on the paragnaths are added to the food mass, and the resulting bolus of food is then pushed into the esophagus by the feeding process of the labrum (Dall et al., 1990; Dall, 1992).

The feeding process described above is relatively efficient when shrimp are handling and ingesting natural food organisms which maintain their structure and integrity. When shrimp are observed feeding on formulated food, however, considerable wastage is evident. This occurs as the food is crushed by the mandibles and manipulated by the mouthparts. Poorlybound pellets may be largely lost to the animal during feeding. While it may be argued that uneaten food contributes material to the nutrient cycle within the pond, this is clearly not the objective of using expensive feeds. Food not consumed is food wasted and subsequent effects on the rearing environment in the pond are unpredictable and may be highly detrimental.

Observations of shrimp feeding in glass aquaria give a useful indication of the quality of shrimp food. Their attraction for shrimp, ease of manipulation, and their stability during handling and ingestion can be readily observed and different feeds compared. Stability in water is a critical factor. Soft, water-soaked pellets are more difficult for shrimp to manipulate. If the pellets disintegrate before, or during feeding, nutrient loss is extremely high. The use of feeding trays (Chapter 8) permits observation of feeding activity and pellet stability over the first critical 1–2 hours after feeding, after which food should normally have been consumed.

After ingestion the food passes down a short tubular esophagus and enters the stomach. This organ in shrimp is divided into two parts: the anterior cardiac stomach, which is a small, thin-walled distensible chamber, and the posterior pyloric stomach.

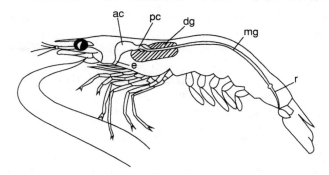

Fig. 2.3 Diagrammatic representation of the alimentary system of a penaeid shrimp. e: esophagus; ac: anterior chamber of proventriculus; pc: posterior chamber of proventriculus; dg: digestive gland; mg: midgut; r: rectum.

The pyloric stomach is reinforced with a complex of calcified teeth and plates, which form the gastric "mill."

Within the stomach the food is first mixed with digestive enzymes secreted by the digestive gland and then ground in the gastric mill. Food particles are size-sorted in the stomach. Larger particles are retained for further processing, while fine particles pass into the digestive gland. Indigestible materials pass from the stomach into the mid- and hindguts prior to excretion. The digestive tract of a penaeid is illustrated in Figure 2.3.

Within the digestive gland the liquefied food is treated with a full range of digestive enzymes: proteases, lipases, and carbohydrates. Emulsifying agents are also present. A range of specialized cells, within the tubular structure of the digestive gland, secrete enzymes (F-cells) and absorb and store nutrients (R-cells and B-cells). Depending on the quality of the food and the balance of nutrients, assimilation efficiencies are generally high for proteins and lipids and lower for carbohydrates.

The digestive process in shrimp is generally rapid and has been reported complete within 4–6 hours at 20°C. The finely divided nature of the food as it enters the digestive gland serves to increase the rate of digestion. The limited storage capacity of the foregut is a significant factor in determining feeding regimens. The volume of the foregut has been estimated at 2–3% of the total body volume. This limited ability to store food, coupled to a rapid digestive process, means that actively growing shrimp must feed more or less continuously.

REFERENCES

BOUJARD, T. and J.F. LEATHERLAND. 1992. Circadian rhythms and feeding times in fishes. Environmental Biology of Fishes **35**:109–131.

CARR, W.E.S. 1982. Chemical stimulation of feeding behavior. In Chemoreception in Fishes, ed. T.J. Hara, pp. 259–273. Elsevier, Amsterdam.

COHEN, D.M. 1970. How many recent fishes are there? Proceedings of the California Academy of Science 38:341–345.

DALL, W., B.J. HILL, P.C ROTHLISBERG and D.C. SHARPLES. 1990. The Biology of the Penaeidae. In Marine Biology, Vol. 27. Academic Press, London.

DALL, W. 1992. Feeding, digestion and assimilation in Penaeidae. In Proceedings of the Aquaculture Nutrition Workshop, Salamander Bay, April 15–17, 1991, eds. G.L. Allan and W. Dall, pp. 57–63. N.S.W. Fisheries, Brackish Water Fish Culture Research Station, Salamander Bay, Australia.

FÄNGE, R. AND D.J. GROVE. 1979. Digestion. In Fish Physiology VIII: Bioenergetics and Growth, eds. W.S. Hoar, D.J. Randall and J.R. Brett, pp. 161–260. Academic Press, New York.

GODDARD, S. 1988. Food and feeding. In Freshwater Crayfish: Biology, Management, and Exploitation, eds. D.M. Holdich and R.S. Lowery, pp. 145–166. Croom-Helm, London.

GROVE, D.J. 1986. Gastro-intestinal physiology: rates of food processing in fish. In Fish Physiology: Recent Advances, eds. S. Nilsson and S. Holmgren, pp. 140–152. Croom-Helm, London.

GROVE, D.J., L. LOZOIDES and J. NOTT. 1978. Satiation amount, frequency of feeding and gastric emptying rate in *Salmo gairdneri*. Journal of Fish Biology 12:507–516.

GWYTHER, D. AND D.J. GROVE. 1981. Gastric emptying in *Limanda limanda* (L.) and the return of appetite. Journal of Fish Biology 18:245–260.

HILL, B.J. and T. J. WASSENBERG, 1985. Feeding behavior of adult tiger prawns, *Penaeus esculentus*, under laboratory conditions. Australian Journal of Marine and Freshwater Research 38: 183–190.

HINDLEY, J.P.R. 1975a. Effects of endogenous and some exogenous factors on the activity of the juvenile banana prawn *Penaeus merguiensis*. Marine Biology 29:1–8.

HINDLEY, J.P.R., 1975b. The detection, location and recognition of food by juvenile banana prawns *Penaeus merguiensis*. Marine Behavior and Physiology 3:193–210.

HINDLEY, J.P.R. and G.C. ALEXANDER. 1978. Structure and function of the chelate pereiopods of the banana prawn. *Penaeus merguiensis*. Marine Biology 48:153–160.

HOLMGREN, S., D.J. GROVE and D.J. FLETCHER. 1983. Digestion and the control of gastro-intestinal motility. In Control Process in Fish Physiology, eds. J.C. Rankin, T.J. Pitcher and R.T. Duggan, pp. 23–40. Croom-Helm, London.

JOBLING, M. 1980. Gastric evacuation in plaice, *Pleuronectes platessa L.* Effect of dietary energy level and food composition. Journal of Fish Biology 17:187–196.l

NEW, M.B. 1995. Status of freshwater prawn farming. Aquaculture Research 26:1–54.

PANDIAN, T.J. and VIVEKANANDAN. 1985. Energetics of feeding and digestion. In Fish Energetics—New Perspectives, eds. P. Tytler and P. Calow, pp. 99–124. Croom-Helm, London.

SEYMOUR, E.A. 1989. Devising optimum feeding regimes and temperatures for the warm water culture of eel, *Anguilla anguilla L.* Aquaculture and Fisheries Management 20:311–323.

SMITH, S.L. 1989. Digestive functions in teleost fishes. In Fish Nutrition, 2d Ed., ed. J.E. Halver, pp. 332–421. Academic Press, New York.

TACON, A.G.J. 1993. Feed ingredients for warmwater fish—Fish meal and other processed feedstuffs. FAO Fisheries Circular No. 856. FAO, Rome.

TACON, A.G.J. 1993. Feed ingredients for crustaceans—Natural foods and processed feedstuffs. FAO Fisheries Circular No. 866. FAO, Rome.

VAHL, O. 1979. An hypothesis on the control of food intake of fish. Aquaculture 17:221–229.

3

\\\\\\

Dietary Requirements

INTRODUCTION

Following the digestion and absorption of food from the digestive tract of fish or shrimp, nutrients become available to meet the metabolic needs of the animal. In the context of feed management some understanding of the nature of essential nutrients used in aquaculture feeds is important when selecting and using feeds of different types. There is extensive literature covering fish and shrimp nutrition and only an outline is included here (reviews are given in Cowey et al., 1985; Halver, 1989; Lovell, 1989; Wilson, 1991; Akiyama et al., 1992; Kaushik and Luquet, 1993; NRC, 1993; Cowey and Wilson, 1994).

While research now extends to a wide range of fish and shrimp species, our knowledge is far from complete for most farmed species. Our understanding of the nutritional requirements of rainbow trout and channel catfish is the most complete. For practical reasons, much of the research on which present feed formulations are based has been conducted on a laboratory scale using juvenile fish. There has been less research into the feed requirements of larger fish and broodstock. This is work which requires large-scale, dedicated facilities, and must extend over repeated generations of fish.

NUTRIENTS

Six major classes of nutrients are found in the feeds of fish and shrimp. These are proteins, lipids, carbohydrates, vitamins, minerals, and water. Of these substances some are used to build and maintain tissues, while others supply energy. Components of proteins, certain types of lipid, minerals, and water are used as structural materials. Carbohydrates, lipids, and protein may be oxidized to provide energy, while trace minerals and water-soluble vitamins function as essential components of coenzymes in biochemical systems. Essential nutrients are those for which fish or shrimp have an absolute dietary requirement and which they are unable to synthesis form other dietary ingredients. Animals fed diets lacking, or containing insufficient levels of available essential nutrients, may exhibit stunted growth, deformity or be prone to disease. Essential nutrients include certain amino acids and fatty acids, most vitamins, and some minerals.

In comparing the nutritional requirements of various cultured fish species the major differences lie in their abilities to utilize carbohydrates, and in their requirements for essential fatty acids (Lall, 1991). There has been little comparative research into the nutritional requirements of the various cultured shrimp species. At present diets are largely formulated on the basis of information pooled from the results of research on the group as a whole.

PROTEIN

Carnivorous fish and shrimp have high dietary requirements for protein. Experiments, which involve feeding animals on high-quality protein in dose-response trials, have shown that carnivorous fish require 40–50% protein in their diets, on a dry weight basis, while omnivorous fish typically require 25–35% dietary protein for maximum growth (Table 3-1). Figure 3.1 illustrates a protein dose-response curve for post-juvenile chinook salmon. A dietary protein level of 44% was considered optimal for practical diets for this species.

Protein requirements vary with the age of the fish or shrimp. Younger animals generally require higher levels of protein to support their more rapid growth than do older animals. Feeds formulated for larvae, fry, and juvenile stages typically contain 5–10% more protein than do grower diets formulated for older, larger fish.

Amino Acids

Proteins are essential ingredients in the diets of fish and shrimp as sources of amino acids. Following digestion, free amino acids are absorbed from the intestine and distributed by the circulatory system to the various organs and tissues. At the tissue level they may variously be metabolized into new proteins for growth and maintenance, or may be used as a source of energy. There are some 20 amino acids, and the biological value of an ingested protein to an animal is determined by the profile of its constituent amino acids, and their availability following digestion and absorption. While animals are able to synthesize and interconvert some of the amino acids, in those fish and shrimp that have been studied in detail, 10 amino acids have been identified as essential dietary ingredients to support maintenance activities and growth. These are the

Table 3–1   Estimated Protein Requirements of Certain Juvenile Fish and Shrimp Species[1]

Species	Protein Requirements % of dry diet	Source
Fish		
Rainbow trout	40–45	Satia, 1974
Atlantic salmon	45	Lall & Bishop, 1977
Tilapia spp	30–35	Jauncey & Ross, 1982
Common carp	30–38	Watanabe, 1988
Milkfish	40	Lim et al., 1979
Channel catfish	25–36	Wilson, 1991
Japanese eel	45	Arai, 1991
Shrimp		
Penaeid spp	38–40	Akiyama, 1992
Kuruma shrimp	50	Deshimaru & Kuroki, 1974
Giant tiger shrimp	30–45	Alava & Lim, 1983
Prawn		
Giant freshwater prawn	40	Millikin et al., 1980

[1]In practical diets, the optimum levels of protein will depend on various factors: fish size, water temperature, feeding rate, the quality of the protein, and the overall digestible energy content of the feed.

amino acids which the animals either cannot synthesize or cannot synthesize in sufficient quantity to support optimal growth. While the same amino acids have been shown to be essential in the diets of all fish and shrimp examined, the quantities required vary between species (Table 3-2).

The amino acid requirements of various fish species have been derived primarily from feeding and growth experiments using chemically defined diets. The use of experimental diets of low energy density, particularly those used in early studies, may have contributed to some of the observed disparities in amino acid requirements, where fish were grown at less than optimal rates. Requirements of different species may be more similar than some published data may indicate (Cowey, 1994).

Nutritionists require amino acid profiles of protein ingredients in order to formulate and manufacture feeds which contain sufficient levels of these essential ingredients. The feeds manufactured for carnivorous species of fish and shrimp generally contain high levels of good quality fish meal, which provide excess levels of the essential amino acids. Feeds containing high levels of proteins of plant or microbial origin, or protein derived from animals other than fish, may require supplementation of essential amino acids to meet the requirements of a particular fish or shrimp species.

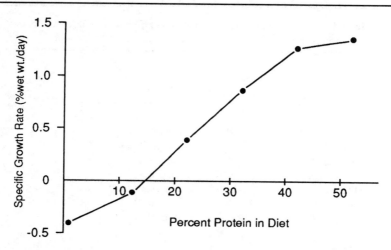

Fig. 3.1 Specific growth rate of post-juvenile chinook salmon, *Oncorhynchus tshawytscha,* fed different levels of protein. From Archdekin et al., 1988

LIPIDS

Dietary lipids are the source of essential fatty acids in aquaculture feeds. They provide a rich source of energy, and dietary phospholipids are vital as structural components of biomembranes. Dietary lipids also serve as carriers for the absorption of other nutrients, including the fat-soluble vitamins A, D, E, and K, and natural or synthetic pigments. Dietary lipids are also the source of essential sterols and phospholipids. Sterols perform a wide range of biological functions and serve as precursors in the synthesis of certain vitamins and hormones.

The lipid nutrition of fish has emerged as a key research area in the development of feeds for aquaculture species. Significant differences in essential fatty acid requirements have been demonstrated both between species and at different lifecycle stages of individual species. For those species, such as the salmonids, where the essential fatty acid requirements have been defined, interest centers on the optimal balance to be achieved between the lipids and other ingredients in terms of their contribution to the overall energy content of the diet. There are also considerable consumer-related interests in the influence of dietary lipids on flesh quality, and in the health aspects associated with their roles in human nutrition (Chapter 9).

Fatty Acids

Over 40 fatty acids are found in nature. Their physical, chemical, and nutritional properties are determined by the number of carbon atoms in the molecule and by the number and positioning of double bonds between carbon atoms. Most naturally occurring fatty acids contain a single carboxyl (COOH) group and a straight, un-

Table 3-2   Quantitative Amino Acid Requirements of Certain Fish and Shrimp Species.

Amino acid	Channel catfish	Japanese eel	Tilapia O. nilotica	Common carp	Penaeid shrimp
Arginine	4.3	4.2	4.2	4.2	5.8
Histidine	1.5	2.1	1.7	2.1	2.1
Isoleucine	2.6	4.1	3.1	2.3	3.5
Leucine	3.5	5.4	3.4	3.4	5.4
Lysine	5.1	5.3	5.1	5.7	5.3
Methionine	—	3.2	—	—	—
(+ cystine)	2.3	5.0	3.2	3.1	3.6
Phenylalanine	—	5.6	—	—	—
(+ tyronsine)	5.0	8.4	5.7	6.5	7.1
Threonine	2.0	4.1	3.6	3.9	3.6
Tryptophan	0.5	1.0	1.0	0.8	0.8
Valine	3.0	4.1	2.8	3.6	4.0
% protein in diet	32.0	38.0	28.0	38.5	36.40
	NRC 1983	Arai 1991	Santiago & Lovell 1988	NRC 1983	Akiyama et al., 1992

branched carbon (C) chain. The carbon chain may contain no double bond (saturated fatty acid), a single double bond (monounsaturated fatty acid), or more than one double bond (polyunsaturated fatty acid [PUFA]). Fatty acids are described and classified using conventional abbreviations based on the carbon chain length, the number of double bonds between carbon atoms (degree of saturation), and by the position of the first double bond in the molecule counting from the terminal methyl group. For example, linolenic acid, an essential fatty acid in the diet of rainbow trout, is described by the expression 18:3n-3. By convention, this describes a fatty acid molecule with 18 carbon atoms, 3 double bonds, and where the first double bond is positioned after the third carbon atom counting from the methyl end of the molecule (n-3). Fats and oils normally found in feedstuffs, and in the fat deposits of most animals, occur in the form of triglycerides. These are esters formed when one molecule of glycerol combines with three similar, or different, fatty acids.

Lipid Requirements

Differences in the lipid requirements of various species of farmed fish and shrimp reflect the natural abundance of the different lipid types found within the food chains of freshwater and marine environments. The predominant fatty acids found in the tissues of fish and shrimp belong to the n-3, or linolenic series. The presence of some n-6, or linoleic series, fatty acids in the tissues of freshwater fish from warm waters, is believed to reflect the consumption of some lipids from terrestrial sources, where fatty acids of the n-6 series predominate. Since fish and shrimp are unable to synthesize fatty acids of the n-3 and n-6 series, these must be provided in their diets.

With the exception of certain carnivorous marine species, fish are able to chain elongate and further desaturate 18:2n-6 (linoleic) or 18:3n-3 (linolenic) fatty acids to their respective highly unsaturated fatty acids (HUFAs), 20:4n-6, 20:5n-3, and 22:6n-3, which are believed to perform the key metabolic roles ascribed to the essential fatty acids.

It has been shown that freshwater fish from cold waters have an essential requirement for fatty acids of the n-3 series, while freshwater fish from warm waters have requirements for either combinations of both n-3 and n-6 fatty acids, or fatty acids of the n-6 series alone. Marine carnivorous species must be provided with highly unsaturated fatty acids, since they lack the ability to chain elongate and further desaturate linolenic acid.

The requirements of shrimp for essential fatty acids have received less study than those of fish. They appear to have requirements for fatty acids of both the lineic and linolenic series. Estimates of the essential fatty acid requirements of various fish and shrimp species are given in Table 3-3.

Crustaceans, including shrimp, have a dietary requirement for cholesterol, which unlike fish, they are unable to synthesize from other nutrients. Cholesterol, a vital precursor of sex and moulting hormones and a constituent of the exoskeleton, is required at dietary levels of 0.5–1.25% of the total diet. Marine invertebrate oils (e.g., squid, shrimp or clam) are rich sources of cholesterol. Shrimp diets are also supplemented with phospholipids. Dietary phospholipids (e.g., 1% lecithin), have been shown to improve both growth and survival in farmed shrimp. While shrimp are able to synthesize phospholipids from other nutrients, the natural rate of biosynthesis appears too low to maintain optimal growth under culture conditions.

Sources

Marine fish oils are rich dietary sources of the n-3 series essential fatty acids, and the diets for most aquaculture species are supplemented with marine oils. For those species with an essential dietary requirement for n-6 fatty acids, diets are generally supplemented with plant oils, which are rich sources of 18:2n-6 fatty acid.

In practice, high levels of oils, 15–20% of the total diet, are included in diets for carnivorous fish species. Oils, in excess of the fishes essential metabolic needs, are provided as sources of energy. This reduces the fishes need to metabolize protein as a source of energy, making more protein available for the growth of new cells and tissues. This effect is termed protein sparing. In formulating diets with high energy content the balance between energy and nutrient content must be maintained. High ratios of energy to nutrients may lead to the excessive deposition of fat in the viscera, or high levels of muscle lipids.

Polyunsaturated fatty acids in marine oils readily undergo auto-oxidation and become rancid, therefore, particular attention has to be paid to the manufacture and storage of diets with high oil content.

Table 3-3   Essential Fatty Acid Requirements of Certain Fish Species

Species	Requirement (% if dry diet)	Source
Rainbow trout	1% 18:3n3	Castell et al. 1972
Channel catfish	1–2% 18:3n3	Satoh et al. 1989
	or 0.5–0.75% n3 HUFA	
Japanese eel	0.5% 18:2n6	Takeuchi et al. 1980
	+0.5% 18:3n3	
Tilapia, *O. nilotica*	0.5% 18:2n6	Takeuchi et al. 1983
Common carp	1% 18:2n6	Takeuchi & Watanabe 1977
	+1% 18:3n6	
Turbot	0.5% n3 HUFA	Gatesoupe et al. 1977
Yellowtail	2% n3 HUFA	Deshimaru & Kuroki 1983
Red sea bream	0.5% n3 HUFA	Yone 1976
	or 0.5% 20:5n3	

CARBOHYDRATES

No essential requirements have been identified for either fish or shrimp for dietary carbohydrates, although carbohydrates, synthesized from dietary proteins and lipids, perform many important functions. They are sources of energy, and components of various biological compounds, including nucleic acids, mucous secretions, and the chitin exoskeleton of shrimp.

The nutritional value of carbohydrates varies among fish. Freshwater and warm-water species are generally able to utilize higher levels of dietary carbohydrates than are cold-water and marine species. The differences may be attributable to the higher intestinal amylase activity found in the former groups (Wilson, 1994).

Carbohydrates, most commonly in the form of starch derived from the endosperms of cereals, are incorporated into aquaculture feeds for various practical reasons. Primarily, the represent a relatively cheap source of energy and their inclusion in feeds may spare part of the more expensive protein for growth. Certain carbohydrates, such as gelatinized starch, cereal glutens, alginates and gums, also function as binding agents in diets, acting to improve water stability. The presence of some carbohydrates may also enhance the texture and palatability of formulated feeds.

VITAMINS

The vitamins are a diverse group of organic compounds that are essential components of fish and shrimp diets, for normal growth, reproduction, and health. Some 15 vitamins have been identified, and most are required by animals in trace amounts, since they are either not synthesized within the animal's body, or are synthesized at too slow a rate to meet the animal's needs. Qualitatively, the vitamin requirements of fish resemble those of mammals. They have been quantified for a number of fish species, and their associated deficiency diseases identified (Halver, 1989).

Vitamins are classified into two main groups, the fat-soluble vitamins and the water-soluble vitamins. Fat-soluble vitamins are absorbed from the digestive tract in association with fat molecules, and can be stored in fat reserves within the body.

Water-soluble vitamins are either used rapidly after absorption, or are broken down and excreted, depending on the animal's needs. Vitamins are subject to denaturing during the feed manufacturing process, in stored feeds, and while immersed in water, before ingestion by fish or shrimp. Amounts above known requirements are generally added to feed mixes in order to compensate for such losses. Correct storage and handling procedures are essential, however, to minimize losses (Chapter 6).

Particular problems may arise in the use of the two antioxidant vitamins, C and E. Vitamin C is prone to oxidation when in contact with water and significant losses may occur when water is added to the feed mix during manufacture. An excess of 50% of vitamin C, in the form of ascorbic acid, may leach from completed feeds during short (10 second) immersion periods in water. Protected forms of vitamin C, such as ascorbyl 2-sulfate and glyceride-coated ascorbic acid offer greater stability during processing and feeding, although significant losses still occur. The levels of vitamin E in finished feeds are affected by the nature and quantity of polyunsaturated fatty acids present and the amounts of other antioxidants. Occurrence of rancid oils will reduce the availability of vitamin E. Oxidation of vitamins is also affected by heat, moisture, pH, and the presence of certain minerals. Relatively little is known about vitamin nutrition in shrimp. The prolonged water immersion times, associated with the slow feeding habits of shrimp, carry the risk of vitamins leaching from feeds. Consequently, feeds are heavily fortified and vitamin supplements typically account for 5–15% of the total feed cost.

There are significant costs associated with analyzing dietary ingredients for their natural vitamin content, which can vary considerably between batches. In view of this, feed manufacturers add excess levels of all of the required vitamins in the form of a premix. Recommended levels are given in Table 3-4.

MINERALS

Fish and shrimp require some 20 inorganic mineral elements in order to maintain health and growth. The essential minerals and trace elements are generally classified as either macro- or microingredients, depending on their concentrations within the bodies of fish or shrimp. There is less information available on the requirements of fish and shrimp for dietary minerals than other feed components. Estimating mineral requirements is complicated by the fact that these animals can obtain minerals not only from their food, but also from water. Minerals are absorbed across the gills of shrimp and across both the gills and skin of fish.

Table 3-4 Vitamin Requirements for Growth of Channel Catfish, Common Carp and Rainbow Trout

Vitamin[1]	Channel catfish	Common carp	Rainbow trout
Vitamin A (IU)	5,500[2]	1,000–20,000	2,000–15,000
Vitamin D (IU)	500–4,000	N.R.[3]	2,400
Vitamin E	50–100	80–300	30–50
Vitamin K	10	N.R.	10
Thiamin	1–20	N.R.	1–12
Riboflavin	9–20	4–10	3–30
Pyridoxine	3–20	4	1–15
Pantothenic acid	10–50	25	10–50
Niacin	14	29	1–150
Folic acid	N.R. or 5	N.R.	0.02
Vitamin B_{12}	0.02	N.R.	0.02
Choline	400	500–4,000	50–3,000
Inositol	N.R.	200–440	200–500
Vitamin C	N.R. or 100	R[4]	100–500

From Lall (1991)

[1] mg/kg of diet unless specified

[2] Values summarized from the published literature

[3] N.R. = not required;

[4] R. = required

Seawater is an excellent source of essential minerals, with the exceptions of phosphorus and iron. Fewer minerals are naturally available to those species cultured in freshwater. Consequently, mineral supplements formulated for freshwater species are more comprehensive than those prepared for marine species. As with vitamins, the known or estimated requirements are generally added to feeds in addition to the natural mineral content of other feed ingredients, of which some, such as fish meal, are significant sources of minerals. A range of pathologies have been identified associated with mineral deficiencies (Tacon, 1992), and the quantitative requirements for a number of species identified (Table 3-5).

ENERGY

Animals require continuous supplies of energy to meet the demands of physical activity and for maintaining body functions. Only when the energy assimilated from the diet exceeds that expended in physical activity and maintenance can growth occur. The capacity of the major feed ingredients to supply energy is, therefore, of prime importance in determining their relative nutritional value.

The metabolic breakdown of dietary carbohydrates, lipids, and proteins leads to the

Table 3-5   Mineral Requirements of Certain Finfish

Mineral	Rainbow trout	Channel carp	Common carp	Japanese eel
Calcium (%)	<0.1	<0.1	<0.1	0.27
Phosphorus (%)[1]	0.7	0.4	0.7	0.3
Magnesium (%)	0.05	0.04	0.05	0.04
Iron (mg/kg)	R[2]	30	—	170
Copper (mg/kg)	3	5	3	—
Manganese (mg/kg)	13	2.4	13	—
Zinc (mg/kg)15–30	15–30	20	15–30	—
Iodine (mg/kg)	R	—	—	—
Selenium (mg/kg)	0.15–0.38	0.25	R	R

From Lall, 1991
[1]Inorganic phosphorus
[2]Required

formation of carbon dioxide, water, and heat, and some of the energy liberated during the aerobic metabolism of these substrates is stored temporarily as either adenosine triphosphate or other related high-energy compounds. These compounds are the sources of energy at the cellular level for metabolic processes such as protein synthesis and osmoregulation, and for mechanical work such as muscular activity during swimming. Heat generated during this process is dissipated across the body of the fish or shrimp and is lost to the water.[1]

Since all forms of energy are convertible into heat energy, the energy content of feeds is conventionally expressed in units of heat.[1] Energy values of the major feed ingredients are determined using bomb-calorimeters, where samples of food are combusted in an atmosphere of oxygen and the released heat energy measured. This gives a direct measure of the gross or total energy value of the food sample. Values for the total and digestible energy contents of feed ingredients are given in Table 3-6.

Since the energy values of the major nutrient types differ, the total energy content of formulated feeds will reflect the relative proportions of dietary ingredients. This balance of energy-contributing nutrients is a critical factor in the formulation of feeds. As described above, energy intake levels are major factors in controlling feeding, hence the absolute amounts of protein, vitamins, and minerals ingested depend to a significant degree on energy intake. The balance of ingredients in formulated diets may therefore

---

[1]The basic unit of heat generally used within the feeds industry is the calorie (cal). This is defined as the amount of heat required to raise the temperature of one gram of water by one degree centigrade. In practice the kilocalorie (kcal), or 1000 calories, is a more convenient measure and is most commonly used. In scientific studies the international unit of energy, the joule (J), is gradually replacing the calorie, where 4.184 = 1 cal.

Table 3-6 Energy Contents of Macronutrients

	Gross energy MJ/kg	Digestible energy MJ/kg
Protein	23.4	16.3
Oil	39.2	33.5
Carbohydrate (uncooked)	17.2	6.5
Carbohydrate (cooked)	17.2	9.6

be regarded as a more critical factor than the absolute levels of specific nutrients (Lall, 1991).

Energy Allocation

Of the energy contained in feeds, a portion becomes available to support growth and essential metabolic functions, while the remainder is lost either as uneaten or undigested food, or as heat and waste metabolites. Figure 3.2 illustrates the patterns of energy transfer that occur. The greatest loss is generally associated with the excretion of undigested food, with smaller losses attributed to the excretion of ammonia and urine.

ME values for feedstuffs are widely applied in the formulation of compound feeds for terrestrial livestock, such as pigs and chickens, since they are relatively easy to determine, and most accurately represent the levels of energy in feeds which are available for growth and maintenance. For fish and shrimp, however, ME values for feedstuffs are more difficult to determine since feces, ammonia, and urine, are excreted into the

Fig. 3.2 The pathways of utilization of food energy.

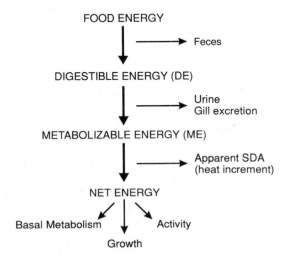

water. Such measurements necessitate the confinement of fish in a metabolism chamber in which they are force-fed and their levels of total fecal, gill, and urinary waste outputs determined. These procedures have not been widely applied, although ME values for some feedstuffs have been reported from experimental work with rainbow trout. In the formulation of compound feeds for fish, DE values are commonly used in place of ME values. These have been determined for both fish and shrimp using a variety of methods to determine the relative energy values of ingested food and feces (Cho et al., 1985). The use of DE values by fish nutritionists is not seen as a major constraint in the formulation of aquaculture feeds, since energy losses from the gills and kidney are small in comparison with the energy losses associated with digestion (Fig. 3.3).

Of the total food consumed by fish and shrimp only a portion becomes available to the animal for growth. Other food energy is lost to the fish as waste products of metabolism, or is used in maintenance functions. At its simplest, growth can be defined as the difference between what enters and what leaves the animal. This is expressed in the balanced energy equation (Winberg, 1956):

$$p \cdot C = M + G \qquad\qquad (3.1)$$

where:
p = proportion of consumed food assimilated
C = food consumed
M = metabolism
G = growth

Fig. 3.3   An energy flow budget in rainbow trout fed a diet containing 20 MJ energy/kg diet. Data from Cho and Kaushik. 1990
F = fecal loss; B+U = branchial and urinary losses; H = heat loss; R = retention in body.

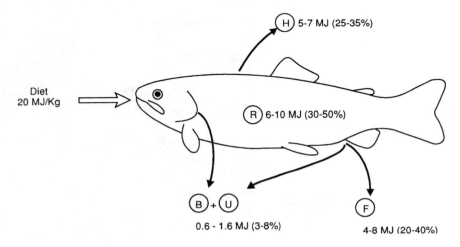

H) 5-7 MJ (25-35%)

Diet 20 MJ/Kg

R) 6-10 MJ (30-50%)

B)+(U)
0.6 - 1.6 MJ (3-8%)

F)
4-8 MJ (20-40%)

Brett and Groves (1979) analyzed data from their own, and many published studies, and calculated general indices of energy use for broadly carnivorous and herbivorous fish species.

$$I = M + G + E \qquad (3.2)$$

where:
I = energy ingested
M = metabolism
G = growth
E = excretion

Carnivorous fish $100I = 44M + 29G + 27E$
Herbivorous fish $100I = 37M + 20G + 43E$

Of interest to fish farmers are the generalized values for the proportion of ingested energy available for growth. Values are typically one third or less of the total ingested. The range of dietary energy ingested between maintenance ration, when no growth occurs, and maximum ration, when maximum growth occurs, represents the scope for growth (an energy budget for rainbow trout is illustrated in Fig. 3.3). There is little comparable data available on the flow of energy and nutrients for shrimp. Figure 3.4 gives a generalized feeding budget for a penaeid shrimp in intensive culture, based on data from studies of diet digestibility and food conversion values.

Fig. 3.4   The fate of feeds released into intensive shrimp ponds. The figures are estimated from diet digestibility and food conversion values. Data from Primavera, 1994

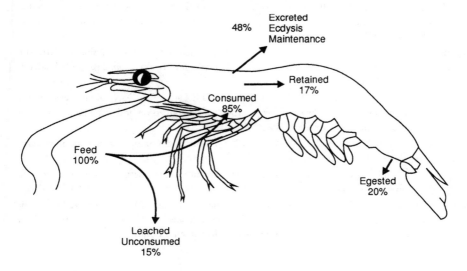

Sex Control

The energy available for growth may contribute to either body (somatic) growth or to the growth (reproductive) of sperm and eggs in sexually mature animals. In some fish species there are significant differences in growth rate between sexes. In some species female fish grow faster than male fish, for example, common carp and eels. In other species, including channel catfish and the tilapias, male fish grow faster than female fish.

The introduction of control methods for sex differentiation, by the administration of sex steroids in early development, have provided effective techniques for the commercial production of monosex populations for certain species (Chourrout, 1987). Techniques for the production of both monosex and sterile populations are now widely applied in fish culture to eliminate the differential growth between sexes, and the deleterious effects of sexual maturation on somatic growth and flesh quality.

REFERENCES

AKIYAMA, D.M. 1992. Future considerations for shrimp nutrition and the aquaculture feed industry. In Proceedings of the Special Session on Shrimp Farming, ed. J. Wyban, pp. 198–205. World Aquaculture Society, Baton Rouge.

AKIYAMA, D.M., W.G. DOMING, and A.L. LAWRENCE. 1992. Penaeid shrimp nutrition. In Marine Shrimp Culture: Principles and Practices, eds. A.W. Fast and L.J. Lester. Elsevier, Amsterdam.

ALAVA, V.R. and C. LIM, 1983. The quantitative dietary protein requirements of *Penaeus monodon* juveniles in a controlled environment. Aquaculture **30**:53–62.

ARAI, S. 1991. Eel, *Anguilla spp.* In Handbook of Nutrient Requirements of Finfish, ed. R.P. Wilson. CRC Press, Boca Raton.

ARCHDEKIN, C.G., D.A. HIGGS, B.A. MCKEOWN, and E. PLISETSKAYA. 1988. Protein requirements of post-juvenile chinook salmon in seawater. Bulletin of the Aquaculture Association of Canada **4**:78–80.

BRETT, J.R. and T.D.D GROVES. 1979. Physiological energetics. In Fish Physiology VIII. Bioenergetics and Growth, eds. W.S. Hoar, D.J. Randal, and J.R. Brett, pp. 270–352. Academic Press, New York.

CASTELL, J.D., R.O. SINNHUBER, J.H. WALES, and D.J. LEE. 1972. Essential fatty acids in the diet of rainbow trout (*Salmo gairdneri*): Growth, feed conversion and some gross deficiency symptoms. Journal of Nutrition **102**:77–86.

CHO, C.Y., C.B. COWEY, and T. WATANABE. 1985. Finfish nutrition in Asia: Methodological approaches to research and development. IDRC, Ottawa.

CHO, C. Y. and S.J. KAUSHIK, 1990. Nutritional energetics in fish: energy and protein utilization in rainbow trout (*Salmo gairdneri*). World Review of Nutrition and Dietetics. **61**:132–172.

CHOURROUT, D. 1987. Genetic manipulations in fish: Review of methods. In Proceedings of the World Symposium on Selection, Hybridisation and Genetic Engineering in Aquaculture Bordeaus, Vol. II, ed. K. Tiews, pp. 111–126. Heenemann, Berlin.

COWEY, C.B., A.M. MACKIE and J.G. BELL (eds.). 1985. Nutrition and Feeding in Fish. Academic Press, London.

COWEY, C. 1988. The nutrition of fish: The developing scene. Nutrition Research Reviews 1:255–280.

COWEY, C.B. 1994. Amino acid requirements of fish: A critical appraisal of present values. Aquaculture 124:1–11.

COWEY, C.B. and R.P. WILSON. 1994. Special issue—Fish nutrition and feeding. Aquaculture 124:1–368.

DESHIMARU, O. and K. KUROKI, 1974. Studies on a purified diet for prawn—I: Basal Composition of the diet. Bulletin of the Japanese Society for Scientific Fisheries. 40:413–419.

DESHIMARU, O. and K. KUROKI. 1983. Studies on the optimum levels of protein and lipid in yellowtail diets. Reports of Kagoshima Prefectural Fishery Experimental Station, pp. 44–79.

GATESOUPE, F. J., C. LEGER, R. METAILLER and P. LUQUET, 1977. Alimentation lipidique du turbot (Scophthalmus maximus L.) 1. Influence du la longueur du chaine des acides grade de la serie w3. Annales d'Hydrobiologie 8:89–97.

HALVER, J.E. 1989. Fish Nutrition, 2d Ed. Academic Press, New York.

JAUNCEY, K. and R. ROSS. 1982. A Guide to Tilapia Feeds and Feeding. Institute of Aquaculture, University of Stirling, Stirling.

KAUSHIK, S.J. and P. LUQUET, (eds.) 1993. Fish Nutrition in Practice. INRA, Paris.

LALL, S.P. 1991. Concepts in the formulation and preparation of a complete fish diet. In Fish Nutrition Research in Asia. Proceedings of the Fourth Asian Fish Nutrition Workshop. Asian Fisheries Society Special Publication 5, ed. S.S. DeSilva, p. 205. Asian Fisheries Society, Manila.

LALL, S.P. and F.J. BISHOP. 1977. Studies on mineral and protein utilization by Atlantic salmon (Salmo salar) grown in sea water. Technical Report No. 688. Fisheries and Marine Service, Environment Canada, Ottawa.

LIM, C., S. SUKHAWONGS and F.P. PASCUAL. 1979. A preliminary study on the protein requirements of milkfish (Chanos chanos) (Forskal) fry in a controlled environment. Aquaculture 17:195–201.

LOVELL, T. 1989. Nutrition and Feeding of Fish. Van Nostrand Reinhold, New York.

MILLIKIN, M.R., A.R. FORTNER, P.H. FAIR, and L.V. SICK. 1980. Influence of dietary protein concentration on growth, feed coversion, and general metabolism of juvenile prawn (Macrobrachium rosenbergii). Proceedings of the World Mariculture Society 9:195.

NRC (NATIONAL RESEARCH COUNCIL). 1983. Nutritient Requirements of Warmwater Fishes and Shellfishes. National Academy Press, Washington, D.C.

NRC (NATIONAL RESEARCH COUNCIL). 1993. Nutrient Requirements of Fish. National Academy Press, Washington, D.C.

PRIMAVERA, J.H. 1994. Environmental and socioeconomic effects of shrimp farming: the Philippine experience. Infofish International 1:44–49.

SANTIAGO, C.B. and R.T. LOVELL. 1988. Amino acid requirements for growth of Nile tilapia. Journal of Nutrition 118:1540–1546.

SATIA, B.P. 1974. Quantitative protein requirements of rainbow trout. Progressive Fish Culturist 36:80–85.

SATOH, S., W.E. POE, and R.P. WILSON, 1989. Effect of dietary n-3 fatty acids on weight gain and liver polar lipid fatty acid composition of fingerling channel catfish. Journal of Nutrition 119: 23–28.

TACON, A.G.J. 1992. Nutritional fish pathology. Morphological signs of nutrient deficiency and toxicity in farmed fish. FAO Fish Technical Paper No. 330, FAO, Rome. p. 75.

TAKEUCHI, T. and T WATANABE. . 1977. Requirements of carp for essential fatty acids. Bulletin of the Japanese Society for Scientific Fisheries 43:541–551.

TAKEUCHI, T., S. SATOH and T. WATANABE. 1983. Requirements of *Tilapia nilotica* for essential fatty acids. Bulletin of the Japanese Society for Scientific Fisheries 49:1127–1134.

TAKEUCHI, T., S. ARAI, T. WATANABE, and Y. SHIMMA. 1980. Requirement of eel *Anguilla japonica* for essential fatty acids. Bulletin of the Japanese Society for Scientific Fisheries 46:345–353.

WATANABE, T. 1988. Nutrition and growth. In Intensive Fish Farming, ed. C.J. Shepherd and N.R. Bromage. BSP Professional Books, London.

WILSON, R.P. 1991 (Ed.). Handbook of Nutrient Requirements of Finfish. CRC Press, Boca Raton.

WILSON, R.P. 1991 Channel Catfish, *Ictalurus punctatus*. In Handbook of Nutrient Requirements of Finfish, ed. R.P. Wilson, CRC Press, Boca Raton.

WILSON, R.P. 1994. Utilization of dietary carbohydrate by fish. Aquaculture 124:67–80.

WINBERG, G.G. 1956. Rate of Metabolism and Food Requirements of Fishes. Nauchnye Trudy Belorusskogo Gos. Univ. Minsk, p. 253 (English translation, Fisheries Research Board of Canada, Transl. No. 194, 1960).

YONE, Y., 1976. Nutritional studies of red sea bream. In Proceedings of the First International Conference on Aquaculture Nutrition. Eds. K.S. Price, W.N. Shaw, and K.S. Danberg. Lewes/Rehoboth, University of Delaware.

4

\\\\\\

Feeding, Temperature, and Water Quality

INTRODUCTION

The interrelationships between water quality and feeding in intensive aquaculture are complex. Water temperatures and dissolved oxygen levels influence feeding activity, metabolism, and growth and hence are of fundamental importance in the determination of both the types and quantities of feeds used. Feeding fish in intensive culture systems in turn leads to the release of potentially harmful organic and inorganic materials into the water. Ammonia, primarily excreted across the gills of fish and shrimp, may exert direct toxic effects on these animals, while phosphorus, nitrogen, and organic solids may affect fish or shrimp indirectly through long-term degradation of their environment. Some harmful compounds are released as excretory products, resulting from normal metabolic activities, while others may be released as a consequence of poor feeding practices or poor food quality.

Nitrogenous wastes, primarily dissolved ammonia, and feces comprising indigestible materials, are produced more or less continuously in actively feeding fish. Ammonia, in its unionized form, is highly toxic to fish and shrimp, while feces released into the water rapidly break up, contributing to both the suspended and settleable solids loading of the water system. In holding systems with adequate water exchange

51

rates, the levels of ammonia and feces are ordinarily controlled by continual dilution by inflowing water. In static pond systems waste materials enter the nitrogen, carbon, and phosphorus cycles within the natural pond ecosystem. Under optimal conditions these compounds are reduced through microbial activity and their elements released into the water. Here they become available as nutrients for algae, and in well-managed pond systems such nutrient cycling makes a valuable contribution to the primary productivity of the pond. The release of farm effluents containing carbon, nitrogen, and phosphorus into receiving waters and surrounding coastal basins may lead to harmful eutrophication or enrichment. Such changes may contribute to conditions that promote the development of algal blooms with their potentially devastating effects on fish and shrimp stocks.

While some parameters of water quality are controllable within routine farm operations, the effective limitations for their control are set during the initial selection of the farm site and in the planning of the farm layout and water supplies.

The effects on water quality of feeding fish in intensive systems has received considerable attention in recent years, particularly in northern Europe and North America, where intensive aquaculture sectors have developed within rigidly enforced regulatory frameworks for the control of water pollution. Initially issues were examined largely in terms of effluent chemistry, but more recently a greater emphasis has been placed on tackling the problems at the source by examining nutritional strategies (Cho et al., 1994). Some significant changes have been made both in the nature of feed ingredients and in the overall nutrient balance of feeds. These changes, coupled to improvements in feeding techniques, have led to significant reduction in pollution loadings in some sectors of the industry, and point the way for others (Table 4-1).

TEMPERATURE

Each species of fish and shrimp has a preferred water temperature range at which feeding, metabolism, and growth are optimal. Fish and shrimp are termed *obligate poikilotherms*, which means that they possess little or no ability to internally regulate their body temperatures independently of their environment. When water temperatures change the fishes body temperature rapidly equilibrates to the new temperature. Heat transfer is by conduction through the body wall and the gills, hence temperature changes occur more rapidly in smaller than in larger animals.

While the body temperatures of fish closely parallels that of their environment they are extremely sensitive to temperature change and some are able to detect temperature changes of less than 0.5°C. Within natural waters fish are able to seek out preferred temperature zones, but in aquaculture systems the temperature ranges available to fish are often restricted. Fish are able to live in habitats with temperatures ranging from −2.5°C – 44°C (Elliot, 1981). No single species can survive over this entire range how-

Table 4-1 Decrease of Nutrient Load Per Tonne of
Salmonids Produced in Denmark

Waste Nutrient	Up to 1980 kg/tonne	Average 1991 kg/tonne
BOD	600	247
Total nitrogen	180	49
Total phosphorus	30	6

Warrer-Hansen, 1993

ever, and each species has a characteristic temperature range with upper and lower
lethal limits. In aquaculture the extremes of temperature tolerance are found in cold-
water and tropical species. Salmonids and certain marine fish species are cultured in
northern temperature regions where part of the production cycle may occur under ice
or at near zero temperatures. In contrast fish and shrimp cultured along the equator
pass through their growth cycle at temperatures exceeding 30°C for most of the year.

Of particular significance in aquaculture is the optimum temperature range for any
cultured species. The optimum temperature range is defined as the range over which
feeding occurs and where there are no signs of abnormal behavior linked to thermal
stress (Elliot, 1981). Within this optimal range it may also be possible to define a nar-
rower growth optimum. Outside the optimal range fish suffer thermal stress and die at
the upper and lower lethal limits (Fig. 4.1). Figure 4.2 contrasts optimal temperature
ranges for a cold-water species, the brown trout *(Salmo trutta)* and a warm-water fish,

Fig. 4.1 Fish growth and temperature. A generalized diagram illustrating the relationship between fish growth and temperature in relation to upper and lower lethal limits.

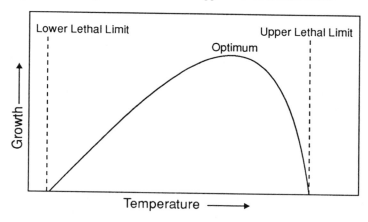

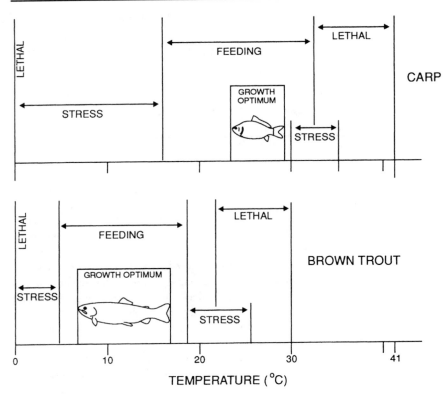

Fig. 4.2 Temperature requirements of the common carp, *Cyprinus carpio,* and the brown trout, *Salmo trutta.* (From Elliot, 1981)

the common carp (*Cyprinus carpio*). Table 4-2 gives figures for optimal temperature ranges for representative species of importance in aquaculture.

As water temperature increases within the fishes' optimal range, there is a corresponding increase in metabolic rate. A result of increasing metabolic rate is a more rapid transit of food through the fishes' digestive system, in turn resulting in increased appetite and food intake. Assuming that food is available to excess, and that dissolved oxygen levels are not limiting, energy intake reaches an optimal level and growth rates increase to a maximum.

OXYGEN

The availability of dissolved oxygen is a critical limiting factor in intensive fish and shrimp culture. Oxygen is a vital requirement of fish, shrimp, and all higher organisms

Table 4-2   Optimal Temperature Ranges for Some Cultured
Fish and Shrimp Species

	Optimal Growth Temperatures (°C)
Rainbow trout	12–18
Atlantic salmon	12–17
Common carp	23–25
Channel catfish	28–30
European eel	18–21.5
Japanese eel	23–30.0
African catfish	25–27.5
Tilapia	25–30
Giant tiger shrimp	28–33
Giant freshwater prawn	25–30

for the production of energy. This is essential to all metabolic functions, including the digestion and assimilation of food, and growth. The requirements of animals vary according to species, age, stage of maturity, and size. Respiratory and blood circulatory systems of fish are adapted to function over a range of dissolved oxygen concentrations. In general terms, in intensive aquaculture, a fundamental aspect of good husbandry is the maintenance of optimal or near optimal dissolved oxygen levels. Where oxygen falls below optimum levels feeding and growth will be impaired and fish or shrimp will be stressed.

While oxygen is a major component of air, comprising one fifth, it is only sparingly soluble in water. Dissolved oxygen levels are influenced by temperature, salinity, and altitude. The levels of dissolved oxygen decrease as temperature, salinity, and altitude increase, and hence are highest at 0°C, zero salinity, and at sea level. Values for dissolved oxygen levels across salinity and temperature ranges are given in Table 4-3. Dissolved

Table 4-3   Solubility of Oxygen in Freshwater and Seawater
at Full Saturation and Normal Atmospheric Pressure

Temperature in °C	Solubility in freshwater in mg/1	Solubility in 35‰ salinity seawater in mg/1
0	14.6	11.3
5	12.8	10.0
10	11.3	9.0
15	10.2	8.1
20	9.2	7.4
25	8.4	6.7
30	7.6	6.1
35	7.1	5.7

oxygen values are generally cited as either milligrams of dissolved oxygen per litre of water (mg/l) or as percentage saturation. Saturation values represent the maximum dissolved oxygen levels possible for any particular temperature, at atmospheric pressure.

Short-term fluctuations occur in dissolved oxygen levels due to a number of factors. DO levels may fall below saturation values or increase to levels of supersaturation. A major factor in the natural fluctuations at farm sites are the respiratory and photosynthetic activities of algae and higher plants. During photosynthesis green plants absorb radiant energy from the sun and convert it into chemical energy through the synthesis of simple carbohydrates. This process can be represented by the chemical equation:

$$6CO_2 \quad + \quad 6H_2O \quad \longrightarrow \quad C_6H_{12}O_6 \quad + \quad 6O_2$$

Carbon dioxide + Water   Glucose   +   Oxygen

Since this process is light dependent, oxygen is produced during daylight hours. At night photosynthesis ceases and plants continue to respire, absorbing oxygen and releasing carbon dioxide. This overall process can produce significant cycles in oxygen and pH (Fig. 4.3). Conditions may fluctuate, depending on plant populations, between levels of subsaturation and supersaturation. In fish farms supplied with flowing water, rivers, or streams, plant activity upstream of the farm will influence DO levels in farm ponds, while similar effects will occur in fish ponds or cages where significant populations of algae are present.

Diurnal fluctuations in DO typically range from 2–3 mg/l. However, in fertile fish ponds, with little or no water interchange fluctuations can be as high as 7–8 mg/l. At levels of this magnitude fish are severely stressed, and growth and food conversion impaired.

Microorganisms active in the decomposition of organic material in the benthos at the bottom of fish ponds, or below fish cages, may also exert a significant demand on DO. This is termed the biochemical oxygen demand (BOD), and is a standard measure taken to determine the effects of organic loadings in aquatic systems.

In intensive shrimp ponds much of the BOD can be traced to the aerobic decomposition of excess feeds and other organic materials in the sediment. Studies have shown that sediments in intensive shrimp ponds consume up to three times as much oxygen as organisms in the water column and the biomass of shrimp combined (Fast et al., 1988).

For any cultured species a knowledge of DO levels and potential fluctuations at the farm site are essential in optimizing feeding regimens. Since oxygen requirements are high, and oxygen has a low solubility in water, occasions may arise where oxygen levels fall close to, or below, the minimum level. Intensive shrimp ponds change from autotrophy, with net positive dissolved oxygen values, to heterotrophy as feed quantities increase during the production cycle.

Under low dissolved oxygen conditions fish and shrimp initially show signs of stress. This will generally first be reflected by fish showing a reduced feeding response and ab-

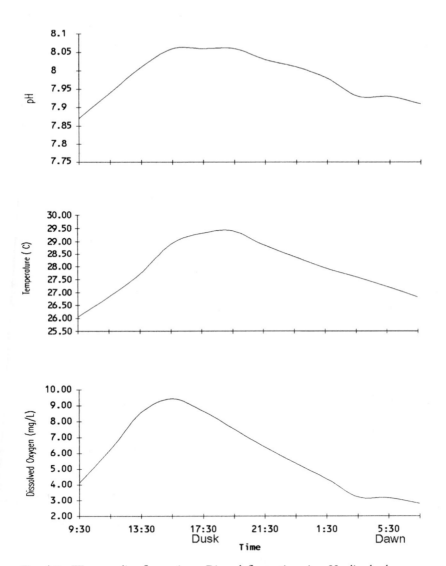

Fig. 4.3 Water quality fluctuations. Diurnal fluctuations in pH, dissolved oxygen, and temperature taken from average values recorded over a 5-day sampling period in a tropical shrimp pond.

normal patterns of swimming and distribution. In extreme cases of oxygen depletion fish may gulp air at the surface. Shrimp respond to conditions of low dissolved oxygen by moving into shallow water or by swimming close to the surface. Shrimp farmers frequently check the margins of shrimp ponds at night, using a flashlight. The presence of shrimp congregating in the shallow margins is taken as an indicator of low dissolved oxygen levels.

While low DO levels are stressful and impair growth, experiments with rainbow trout (Smart, 1981) and channel catfish (Carter and Allen, 1976) have indicated that maintaining dissolved oxygen levels above the required minimum did not increase food intake, food conversion efficiency, or growth rates. However, fluctuations in dissolved oxygen levels, above the minimum requirement, were shown to have an adverse effect on food conversion values (Smart, 1981).

OXYGEN REQUIREMENTS AND FEEDING

Fish and shrimp require energy in order to move, to find and digest food, to grow, and for maintenance of body functions. Energy stored in their food must be converted, through metabolic processes, to release energy. Oxygen, along with an organic substrate, is needed for all oxidative metabolic processes. Such oxidative (aerobic) pathways are dominant in organisms which have a fairly reliable oxygen source and are biochemically more efficient than anaerobic pathways. The amount of oxygen which a fish requires for these processes over a period of time is called its oxygen consumption rate. Oxygen consumption rates are affected by a number of factors, including body weight, water temperature, and level of activity. In general larger fish use more oxygen per hour than smaller fish, although per unit body weight, smaller fish use more oxygen than larger fish. At higher temperatures and activity levels fish require more oxygen than at lower temperatures or resting levels of activity.

Fish species show marked adaptation in their ability to live and function in waters with different ranges of dissolved oxygen levels. Warm water species are generally able to thrive in waters with lower DO levels than are cold water species. For example, salmonids have high DO requirements, with a minimum tolerance level of 5–6 mg/l. This requirement reflects their natural habitat of temperate upland streams. In contrast the tilapias have lower overall requirements reflecting their natural habitat in slow-moving tropical rivers, lakes, and ponds. Tilapia require a minimum dissolved oxygen level of 3 mg/l.

Fish exchange respiratory gases across the surface of the gill lamellae. Oxygen is absorbed while carbon dioxide is excreted. Transfer of gases is by diffusion across the thin membranes of the lamellae which separate the blood circulatory system from the flow of water through the gills. The oxygen content of the blood depends on a number of factors including the partial pressures of oxygen and carbon dioxide in the water, pH, temperature, and the fishes' activity level.

Low DO levels are a common cause of mortality and reduced growth in intensive shrimp ponds. Dissolved oxygen tolerance values for penaeid shrimp range from 1.2–2.2 mg/l (Primavera, 1993), while dissolved oxygen values below 0.9 mg/l are lethal for most species.

FEEDING AND METABOLIC RATE

Following feeding there is an increase in the rate of oxygen consumption. Shortly after feeding the rate of oxygen consumption rises to a peak, and then gradually falls back to the prefeeding resting level (Fig. 4.4). An increase in metabolic rate following the ingestion of food is well documented and has been termed the specific dynamic action (SDA) and relates to the metabolic activities associated with the assimilation and absorption of nutrients. The exact causes are not clear although the effects and a range of factors which influence the SDA in fish have been reported (Jobling, 1981). In extreme cases at high temperatures, when dissolved oxygen levels in water are reduced and resting metabolic oxygen demand by fish is high, feeding fish to satiation can result in such high oxygen demand, necessary to satisfy SDA, that fish mortalities can result (Roberts and Bullock, 1989). Under normal conditions the maximum rates of oxygen consumption induced by feeding are approximately double those of resting fish.

The duration of the effect of SDA on oxygen consumption varies with a number of factors, including water temperature, meal size, food composition, and fish weight. Peak oxygen demand generally occurs within 12 hours after ingestion of the meal and some evidence suggests that the duration of the effect may be linked to the rate of passage of food through the gut. SDA may also be linked to appetite and the control of food intake.

Fig. 4.4   Food ingestion and oxygen consumption. The shaded area below the curve represents the increased oxygen demand which follow feeding. (From Jobling, 1981)

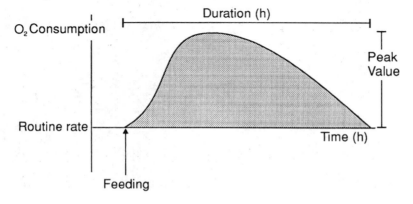

An appreciation of SDA is of particular importance to fish and shrimp farmers when dissolved oxygen levels fluctuate with the diurnal cycles of plant photosynthetic activity. Feeding fish must be scheduled to avoid maximum oxygen demand, resulting in part from SDA, coinciding with low ambient oxygen levels. This may necessitate feeding fish earlier, or omitting midday or evening feeds, in order to avoid maximum oxygen demand occurring around dawn, when dissolved oxygen levels are lowest.

OXYGEN DEPLETION

Routine monitoring of dissolved oxygen (DO) levels is an essential component of feed management. Actual or predicted fluctuations in DO levels generally necessitate some adjustment of feeding rates, or in extreme circumstances cessation of feeding, in order to alleviate stressful and potentially harmful conditions. DO should be measured routinely on a daily basis, and measurements taken at the surface, at mid-depth and near the bottom of the culture system in order to establish an oxygen profile. Chemical methods, based on the Winkler test, are readily available as simple-to-use kits, designed for use on fish farms. Electronic DO meters are also widely available and can be pur-

Fig. 4.5   Water quality monitoring equipment. A multifunctional water quality meter and probe. This equipment is designed to measure and record simultaneously a range of water quality parameters, including salinity, pH, dissolved oxygen, and temperature.

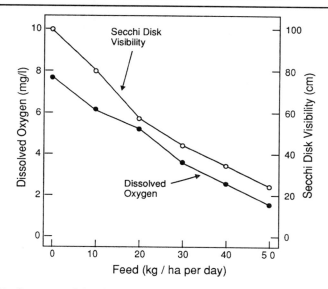

Fig. 4.6 Feeding rate and dissolved oxygen concentrations. The effects of increased feeding levels on dissolved oxygen concentrations at dawn, and Secchi disk visibility in catfish production ponds (From Boyd, 1991).

chased with a range of functions for the simultaneous measurement of other water quality parameters, such as temperature, salinity, and pH. Combined with data loggers, electronic meters and probes can be used to provide farmers with continuous recordings of DO, and other water quality measurements (Fig. 4.5).

In temperate climates most problems associated with DO occur during the warmer summer months, and result from a combination of factors. At higher temperatures the oxygen holding capacity of water is reduced, and this typically coincides with periods of active growth, high stocking densities, and elevated feeding levels. The rapid growth of microorganisms, algae, and higher plants during the warmer months also exerts additional oxygen demand. A combination of any, or all, of these factors may result in critically low dissolved oxygen levels (Fig. 4.6). In the tropics and subtropics, these factors are present year-round.

Where diurnal DO fluctuations are caused by the photosynthetic-respiratory activities of plants, DO levels should be measured at daybreak and monitored until safe levels are reached to commence feeding, taking into account the fact that the oxygen requirements of fish peak several hours after feeding. Food consumption should be carefully monitored under these conditions and feeding rates adjusted as necessary in order to avoid wastage and to prevent any additional contribution to the oxygen demands.

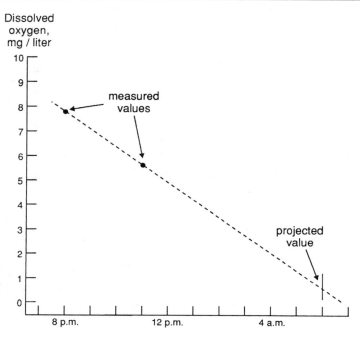

Fig. 4.7   Night time dissolved oxygen depletion. A graphical method for predicting the nighttime decline of dissolved oxygen concentration in fish ponds. (From Boyd, 1991)

Various procedures can be used to predict nighttime falls in DO level (Boyd et al., 1978). On a plot of DO vs. time, two points can be determined, one at dusk and one several hours later, and a straight line drawn through the two points used to predict the lowest levels which will occur just before dawn (Boyd, 1991) (Fig. 4.7).

WATER EXCHANGE AND AERATION

Periods of oxygen depletion occur more commonly in culture ponds than in tanks, raceways, or cages. Low water exchange rates, photosynthesis in plants, and the respiratory activity of the pond biomass create fluctuations in DO levels. In tank and raceway systems where high stocking densities are supported by high rates of water exchange, oxygen depletion most commonly occurs as a result of a systems failure in the water supply or aeration system, or contamination of the water supply. Water exchange through cages or pens is generated by currents, wind, and the movements of the fish. Fluctuations in DO levels may arise as a result of the photosynthetic activity of plants or in extreme cases as a result of phytoplankton blooms.

Fig. 4.8 A paddlewheel aerator. This is the most commonly used equipment for supplementing dissolved oxygen levels in shrimp and warm-water fish ponds.

Oxygen depletion can be alleviated by water exchange, the use of air diffusers, mechanical aerators (Fig. 4.8), or by the use of pure oxygen aeration systems (Fig. 4.9) (Boyd and Watten, 1989; Colt and Orwicz, 1991). Early awareness or anticipation of problems is a critical element in alleviating or avoiding harmful conditions. Careful monitoring of dissolved oxygen levels, combined with observations of the behavior of fish or shrimp, is essential and should be routine activity in the operations of the farm. Nighttime aeration in channel catfish ponds has been shown to be effective in reducing oxygen depletion, and has been shown to result in higher fish yields (Fig. 4.10).

FISH FARM WASTES

Wastes produced during the farming of fish and shrimp fall into two categories: solid wastes derived from uneaten food, dust and feces, and soluble excretory products, primarily ammonia and urine, dissolved organic materials, and carbon dioxide (Fig. 4.11). These materials are potentially harmful to both the immediate farm environment and waters receiving farm effluent.

Fig. 4.9   Liquid oxygen. Pure oxygen is commonly stored in bulk for use in hatcheries, and intensive fish production facilities.

Uneaten Food

While considerable progress is being made by the manufacturers of fish and shrimp feeds to produce highly palatable and digestible foods, waste products are an inevitable part of the feeding process and must be accounted for in farm management procedures. High levels of wasted uneaten food can often be attributed to poor feeding practices and inaccurate estimates of the biomass of animals being fed. Levels of uneaten food have been shown to differ in accordance with the type of food used, that is, wet, moist, or dry. Estimates of uneaten food from pond and tank culture of rainbow trout in Denmark gave the percentage of unconsumed food as trash fish, 10–30%, moist pellets, 5–10%, and dry feeds, 1–5% (Warren-Hansen, 1982). The significance of these different waste levels, estimated for various food types, is reduced if wastage is accounted for in terms of dry matter.

At cage sites, estimates of food wastage are higher than those for ponds, tanks, or raceways. Detailed studies in Scotland estimated wastage from rainbow trout and salmon farms, due to uneaten food, at values up to 20% of food fed. This is a high figure, significantly affecting the profitability of rearing fish in cages. In response to the economic and environmental problems of food wastage at cage sites, waste detection

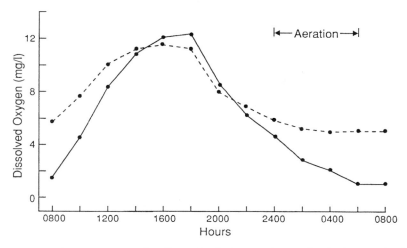

Fig. 4.10 Nighttime aeration. The effects of nighttime aeration on dissolved oxygen levels in catfish ponds: aerated pond, dotted line: unaerated pond, solid line. (Boyd, 1991)

systems are being developed in an attempt to eliminate or reduce the problems associated with overfeeding (Chapter 7).

There have been few studies of formulated food consumption by shrimp in ponds. Accurate determination of levels of uneaten food is complicated by the nature of the shrimp pond environment and the availability of natural food organisms. Fifteen percent was given as an estimate of uneaten food for giant tiger shrimp farms in the Philippines (Primavera, 1993). A detailed study conducted on karuma shrimp ponds in France indicated much higher levels of food wastage, which were related to the population densities of preferred natural food organisms (Reymond and Lagardère, 1990).

The fate of uneaten food depends on the type of holding system. In static ponds with low water exchange rates most uneaten food remains in the pond. Feeds used in cage rearing of fish and shrimp are generally of relatively high density, and settle rapidly below the cage if not consumed. Among the sediment the food pellets break apart and may be consumed by other animals, such as polychaete worms, molluscs, and insects. Waste food not consumed is subject to bacterial breakdown. Under aerobic conditions this will create an additional oxygen demand. If oxygen supplies become depleted the sediment will become anaerobic and hydrogen sulfide, produced by heterotrophic bacteria, may accumulate to harmful levels. Hydrogen sulfide is a highly soluble gas, which is toxic to fish and shrimp at very low concentrations, and it can pose significant problems under fish cages, or in culture ponds, where there is insufficient water exchange for dilution to safe levels.

In sediment the oxidation reduction potential (redox) controls the reduction of sulfates to sulfides. In water, hydrogen sulfide exists in unionized (H_2S) and ionized forms

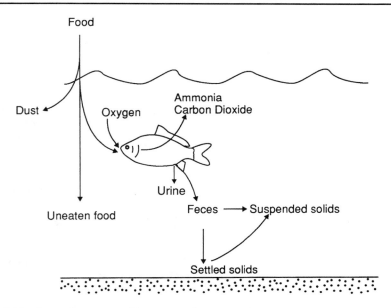

Fig. 4.11   Aquaculture wastes. The sources of organic and inorganic wastes in intensive aquaculture.

(HS⁻ and S₂⁻); of these the unionized form is considered toxic to aquatic animals. The concentration of unionized H_2S is dependent on pH, temperature, and salinity. The percentage of unionized H_2S decreases as pH increases. In general, hydrogen sulfide accumulates in sediment that is highly reduced (redox potential <150mV), within the pH range 6.5–8.5, and which is low in iron (Jakobsen et al., 1981). Hydrogen sulfide is readily oxidized and is seldom observed in surface water. It may be present at toxic levels within 1–2 cm of the sediment surface and may pose significant problems for bottom-dwelling animals such as shrimp. Recommended maximum safe levels for continuous exposure to hydrogen sulfide are 0.005 mg/l for shrimp (Chien, 1992) and 0.002 mg/l for freshwater fish (Boyd, 1982).

Disturbance of anaerobic sediments by aeration devices, or during netting operations in ponds, may release harmful amounts of the gas. At cage sites, excessive water movements resulting from weather disturbances or strong tidal action, may also disturb sediments. Less-soluble methane gas is also released from anaerobic sediments and may be visible in ponds or at cage sites as it bubbles to the surface.

Feces

Up to one third of the content of feeds used in intensive aquaculture may be indigestible. This fraction, combined with mucus, bacteria, and cells shed from the walls of the digestive tract, is excreted as feces. Feces may contain up to 30% of the dietary carbon consumed and 10% of consumed nitrogen. This results in large quantities of waste organic materials entering the farm environment. Feces may either settle to the sediment or break up, increasing the levels of suspended solids in the water column. Dust particles from poorly bound feeds will also contribute to the suspended solids content of water. High levels of suspended solids are potentially damaging to the delicate gills of fish and shrimp. They may cause physical injury, which increases the risk of bacterial infection, or may physically block the flow of water across the gill surface, impairing respiratory function.

This kind of pollution loading is particularly significant in intensive fish farming, with heavy feeding and high stocking densities. Stocking and feeding policies should therefore be controlled in accordance with the tolerance of the species to suspended solids. Fishes show a range of tolerance; values of 25ppm are generally regarded as safe. At levels between 25–88 ppm gill structures may be protected, depending on the nature of the solids. Most species have poor protection from suspended solids at levels exceeding 80 ppm.

Solid waste amounts, produced during the culture of fish have been studied most extensively in the freshwater production of rainbow trout. Values estimated from cage, pond, and tank rearing systems range from 0.52 to 0.65 tonnes of solid waste for the annual production of each tonne of fish (Sumari, 1982; Warren-Hansen, 1982; Merican and Phillips, 1985). These are mean values of data collected from a number of farms. The results from different farms varied considerably, reflecting the different feeding levels and management practices employed on the farms that were surveyed.

Mass balance calculations have also been used to estimate the volumes of uneaten food, feces, and soluble wastes. Using this approach, the total loadings of waste nutrients are determined from the difference between the quantities of nutrients fed and the quantities retained in fish or shrimp carcasses. Estimates of the nitrogen balance for tilapia indicate that 98 kg of nitrogen are lost to the environment for each tonne of tilapia produced (Beveridge and Phillips, 1993). This estimate is based on the assumption that the nitrogen content of tilapia is 3% and that of a particular tilapia diet is 8%. At an FCR of 1.6:1, and estimating that 20% of food used is uneaten, then 102.4 kg of nitrogen is consumed but only 30 kg is retained per tonne of tilapia produced. It is further estimated that 360 g of feces are produced per kg of food ingested, and that if fecal nitrogen content is 4%, then 18.4 kg of nitrogen are voided as feces while the balance of 54 kg is excreted as ammonia and urine. This mass balance estimate is illustrated in Figure 4.12. Nitrogen and phosphorus budgets have been similarly calculated for shrimp. Estimates of wastage of nitrogen and phosphorus in intensive giant tiger shrimp suggest total nitrogen loads of 57.3–118.1 kg and total phosphorus loads of 13.0–24.4 kg per tonne of shrimp production, at food conversion ratios of 1.2:1 and 2.0:1, respectively (Phillips et al., 1993).

Problems associated with the accumulation of uneaten food and feces are most acute in static ponds and at cage and enclosure sites which do not receive adequate flushing. The occurrence of bubbles of methane gas, or the characteristic "rotten eggs"

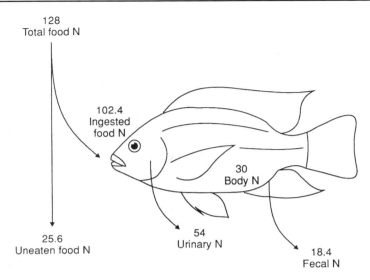

128
Total food N

102.4
Ingested
food N

30
Body N

25.6
Uneaten food N

54
Urinary N

18.4
Fecal N

Fig. 4.12 Mass balance. A mass balance estimation for nitrogen (N), based on the production of 1 tonne of tilapia. (From Beveridge and Phillips, 1993)

smell of hydrogen sulfide gas, are indicators that anoxic chemical changes are occurring and remedial action is necessary.

Under these conditions it may be necessary to relocate the cages until the chemistry and fauna of the sediment revert to their original condition. Rotation of cages is disruptive to fish culture operations, and sites may take many years to recover, particularly in colder regions. Since the problems are largely associated with overfeeding, they call for critical examination of both feeding policy and fish or shrimp stocking densities.

Harmful materials in organic sediments deposited in semi-intensive or intensive fish and shrimp ponds typically approach critical levels toward the end of each production cycle, when stocking densities and feeding levels are highest. In addition to uneaten food and feces, fertilizers, plankton, and solids carried in the water supply, also contribute to sediment buildup. The deterioration of pond water quality, and accumulation of harmful sediments is most commonly attributable, however, to excessive use of feeds (Boyd, 1992).

Treatments of pond bottoms can only be carried out effectively between production cycles, and typically involve a combination of flushing, drying, and removal of excess or anaerobic solids (Fig. 4.13). Ponds must be flushed immediately after harvesting before the sediments consolidate. Rakes, drag-chains or hydraulic jets are often used to further loosen and flush sediments. In intensive culture ponds it may be necessary to remove or replace the top, 5–20cm, reduced layer of sediment, either manually using dredges, or mechanically using earthmoving equipment.

Fig. 4.13   Soil Treatment. A shrimp pond drained between production cycles. Excess or anaerobic solids can be removed if necessary once the pond is dry and the bottom consolidated.

Separation of Solid Wastes

Solids in fish farm effluents can be separated by a number of methods including the use of microscreens, swirl separators, or sedimentation. Most commonly the effluent is passed though shallow settlement ponds designed to allow sufficient retention times for solids to settle to the bottom (Fig. 4.14). These ponds can be drained at intervals and the separated solids removed. Since the solids are rich in nitrogen and phosphorus they may have value as agricultural fertilizer.

Ammonia

About 40–90% of the nitrogenous waste, resulting from the metabolism of proteins, is excreted by fish and shrimp as ammonia from the gills, and in urine. Most of this ammonia is excreted passively in the unionized form from the gills. On release the ammonia molecules react in water to form ammonium hydroxide, which readily dissociates to ammonium and hydroxyl ions. The total ammonia dissolved in water is therefore present in two forms: ionized ammonia (NH_4^+) and unionized or free ammonia (NH_3). The relative proportion of the two forms depends primarily on pH and to a lesser extent on temperature and salinity. Accordingly the temperature and pH of the

Fig. 4.14 Settlement ponds. Settlement ponds for the removal of solids situated below a freshwater farm producing Atlantic salmon smolt. The retention times for farm effluent within the shallow ponds are calculated to permit the majority of solids to settle within a single pass.

water should always be measured at the same time that samples are taken for ammonia determinations.

Ammonia Toxicity

The significance of ammonia for fish culture is the acute toxicity of free ammonia to fish. As ammonia concentrations in water increase, excretory processes are impaired, and ammonia levels build up in the animals tissues and circulatory system. There is a resultant elevation of blood pH and adverse effects on enzyme-catalyzed reactions and the stability of cell membranes. High ammonia levels increase oxygen consumption, cause gill damage, and reduces the ability of blood to transport oxygen (Chien, 1992).

Long-term exposure of fish and shrimp to unionized ammonia can result in loss of appetite, poor growth, and in extreme cases, death. The maximum acceptable limit for fish varies between species. For salmonids a level of 0.02 mg/l is the maximum limit which does not appear to impair feeding and growth. Tolerance levels of tilapia and channel catfish are higher at 0.12 mg/l. For penaeid shrimp safe levels of 0.1 mg/l are generally accepted (Chien, 1992). In intensive aquaculture ammonia levels should be routinely determined and care taken to distinguish between total ammonia levels, as determined using conventional test-kits or meters, and the percentage of unionized

Table 4-4 Percentage of Total Dissolved Ammonia Which is Unionized (Toxic Form) as a Function of pH and Water Temperature

pH	Water Temperature				
	5°C	10°C	15°C	20°C	25°C
6.5	0.04	0.06	0.09	0.13	0.18
6.7	0.06	0.09	0.14	0.20	0.28
7.0	0.12	0.19	0.27	0.40	0.55
7.3	0.25	0.37	0.54	0.79	1.10
7.5	0.39	0.59	0.85	1.25	1.73
7.7	0.62	0.92	1.35	1.96	2.72
8.0	1.22	1.82	2.65	3.83	5.28

ammonia present. This necessitates taking pH and temperature readings and referring to published tables (Table 4-4).

Where ammonia levels are high it may be necessary to flush more water through the system or if this is not possible, feeding levels should be temporarily reduced. High ammonia levels are most likely to occur in pond systems with low water turnover, or in systems where water is being recycled. At farm sites using tanks, raceways, or cages, water exchange rates sufficient to maintain adequate oxygenation will dilute ammonia to within safe limits.

Phosphorus

Fish and shrimp have essential requirements for dietary phosphorus. It has a range of metabolic functions and is required in fish for normal growth and bone development. Food is the main source of phosphorus since, unlike most other minerals, na-

Table 4-5 Maximum Permissible Levels of Discharge of Pollutants from Danish Fish Farms

Waste	Level mg/1
Suspended solids (SS)	1.0
Organic matter (BOD)	3.00
Phosphorus (P)	0.05
Ammonia (NH_3)	0.40
Total nitrogen (N)	0.60

(Dissolved oxygen in outlet water $\geq$ 60% saturated)

From Jensen, 1991

Table 4-6   Regulatory Requirements for Feeds Used on Danish Fish Farms

	Maximum FCR	Gross energy Mcal/Kg/ dry diet	Minimum digestibility %	Protein maximum in food %	Phosphorus maximum in food %	Dust Maximum in food %
1989	1.2	5.6	70	50	1.0	1.0
1990	1.1	5.7	74	50	1.0	1.0
1992	1.0	6.0	78	45	0.9	1.0

From Jensen, 1991

tural phosphate concentrations are low in both freshwater and seawater. Diets deficient in this element can suppress appetite and growth, and affect normal bone formation.

Problems associated with the release of phosphorus in fish farm effluents are widely recognized. Phosphorus discharged into the environment, in uneaten food and feces, stimulates the growth of algae, and contributes to the process of eutrophication. In consequence feed ingredients should be selected with high phosphorus bioavailability, and levels of dietary phosphorus in excess of the fish's requirements should be avoided. The availability of dietary phosphorus to fish varies depending on the nature of its source and the species of fish. Phosphorus from more soluble forms, such as monocalcium or dicalcium phosphates is generally more available than phosphorus from complex salts. Phytin phosphorus from plants has a low availability to most fish. Fish, such as carp and tilapia, with limited acidic gastric secretions, assimilate less phosphorus from fish meal than do salmonid species.

CONTROL OF WASTES

In addition to their effects on animals within culture systems, excretory and waste products, can have wide-ranging effects beyond the immediate farm environment, on those waters receiving farm effluent. This is a critical issue in feed management since effluent quality can be linked directly to feeds and feeding practices and is regulated within water pollution control laws in many countries. The regulations that govern effluents from aquaculture operations are subject to periodic review as sectors of the industry develop and more becomes known about the physical and chemical natures of fish farm effluents and their effects on receiving waters. In most countries aquaculture is regulated within the framework of existing laws, set in place to control industrial and agricultural effluents.

The main pollutants associated with fish farming are phosphorus (P), nitrogen (N), organic matter with its associated biochemical oxygen demand (BOD), and suspended solids (SS). In those countries with extensive pollution controls, limits are generally set on the discharge of pollutants from fish farms. Table 4-5 shows the maximum permissible discharge of pollutants from freshwater fish farms in Denmark, and which are typical of current European standards (Rosenthal et al., 1993).

In addition to discharge controls, some countries have recently introduced legislation which requires that all fish feeds have a certain minimum digestibility and a maximum food conversion ratio (FCR). Table 4-6 lists the regulatory requirements for feeds used in Denmark. In Norway, regulations introduced in 1993 have set maximum FCRs at <1.2:1 at salmon sea cage sites, and <1.1:1 at freshwater smolt production farms (Leffertstra, 1993). Legislation of this nature places the onus on the feed manufacturers to use high-quality, highly digestible raw materials, and on the farmers to carefully select and store their feeds and to avoid any wasteful feeding practices.

REFERENCES

BEVERIDGE, M.C.M. and M.J. PHILLIPS. 1993. Environmental impact of tropical inland aquaculture. In Environment and Aquaculture in Developing Countries, eds. R.S.V. Pullin, H. Rosenthal, and J.L. Maclean, pp. 213–236. International Center for Living Aquatic Resources Management (ICLARM), Manila.

BOYD, C.E. 1982. Water Quality Management for Pond Fish Culture. Elsevier Scientific Publishing Company, Amsterdam.

BOYD, C.E. 1991. Empirical modeling of phytoplankton growth and oxygen production in aquaculture ponds. In Aquaculture and Water Quality, eds. D.E. Brune and J.R. Tomasso, pp. 363–395. World Aquaculture Society, Baton Rouge.

BOYD, C.E. 1992. Shrimp pond bottom soil and sediment management. In Proceedings of the Special Session on Shrimp Farming, ed. J. Wyban, pp. 166–181. World Aquaculture Society, Baton Rouge.

BOYD, C.E., R.P. ROMAIRE and E. JOHNSTON. 1978. Predicting early morning dissolved oxygen concentrations in channel catfish ponds. Transactions of the American Fisheries Society 107:484–492.

BOYD, C.E. and B.J. WATTEN. 1989. Aeration systems in aquaculture. Aquatic Sciences 1(3):425–472.

CARTER, R.R. and K.O. ALLEN. 1976. Effects of flow rate and aeration on survival and growth of channel catfish in circular tanks. Progressive Fish Culturist 38:204–206.

CHIEN, Y.H. 1992. Water quality requirements and management for marine shrimp culture. In Proceedings of the Species Session on Shrimp Farming, ed. J. Wyban, pp. 144–156. World Aquaculture Society, Baton Rouge.

CHO, C.Y., J.D. HYNES, K.R. WOOD and H.K. YOSHIDA. 1994. Development of high-nutrient dense, low pollution diets and prediction of aquaculture wastes using biological approaches. Aquaculture 124:293–305.

COLT, J. and C. ORWICZ. 1991. Aeration in intensive aquaculture. In Aquaculture and Water Quality, eds. D.E. Brune and J.R. Tomasso, pp. 198–271. World Aquaculture Society, Baton Rouge.

ELLIOTT, J.M. 1981. Thermal stress on freshwater teleosts. In Stress and Fish, ed. A.D. Pickering, pp. 209–245. Academic Press, London and New York.

FAST, A.W., K.E. CARPENTER, V.L. ESTILO and H.J. GONZALES. 1988. Effects of pond depth and artificial mixing on dynamics of Phillipine brackishwater shrimp ponds. Aquaculture Engineering 7:349–361.

JAKOBSEN, P., W.H. PATRIDE and B.G. WILLIAMS. 1981. Sulfide and methane formation in soils and sediments. Soil Science 132(4):279–287.

JENSEN, P. 1991. Aquafeeds: reducing environment impact. Feed International, **August:** 6–12.

JOBLING, M. 1981. The influence of feeding on the metabolic rate of fishes: A short review. Journal of Fish Biology 18:385–400.

LEFFERTSTRA, H. 1993. Regulating effluents and wastes from aquaculture production in Norway. In Workshop on Fish Farm Effluents and their Control in EC Countries, eds. H. Rosenthal, V. Hilge, and A. Kamstra. Institute for Marine Science, Kiel.

MERICAN, Z.O. and M.J. PHILLIPS. 1985. Solid waste production from rainbow trout *Salmo gairdneri* Richardson, cage culture. Aquaculture and Fisheries Management 16:55–70.

PHILLIPS, M.J., C. KWEI LIN and M.C.M. BEVERIDGE. 1993. Shrimp culture and the environment: Lessons from the world's most rapidly expanding warmwater aquaculture sector. In Environment and Aquaculture in Developing Countries, eds. R.S.V. Pullin, H. Rosenthal, and J.L. Maclean, pp. 171–197. International Centre for Living Aquatic Resources Management (ICLARM), Manila.

PRIMAVERA, J.H. 1993. A critical review of shrimp pond culture in the Philippines. Reviews in Fisheries Science 1:151–201.

REYMOND, H. and J.P. LAGARDÈRE. 1990. Feeding rhythms and food of *Penaeus japonicus* Bate (Crustacea, Penaeidae) in salt marsh ponds: Role of halophilic entomofauna. Aquaculture 84:125–143.

ROBERTS, R.J. and A.M. BULLOCK. 1989. Nutritional pathology. In Fish Nutrition (2d Ed.), ed. J.E. Halver. Academic Press, San Diego.

ROSENTHAL, H., V. HILGE and A. KAMSTRA. 1993. Workshop on Fish Farm Effluents and their Control in EC Countries. Institute for Marine Science, Kiel.

SMART, G.R. 1981. Aspects of water quality producing stress in intensive fish culture. In Stress and Fish, ed. A.D. Pickering, pp. 277–293. Academic Press, London.

SUMARI, O. 1982. A report on fish farm effluents in Finland. In Report of the EIFAC Workshop on Fish Farm Effluents, ed. J.S. Alabaster. EIFAC Technical Report 41:21–27.

WARRER-HANSEN, I. 1982. Methods of treatment of waste water from trout farming. In Report of the EIFAC Workshop on Fish Farm Effluents, ed. J.S. Alabaster. EIFAC Technical Report 41:113–121.

WARRER–HANSEN I. 1993. Fish farm effluents and their control in EC countries. In Workshop on Fish Farm Effluents and their Control in EC Countries, eds. H. Rosenthal, V. Hilge, and A. Kamstra. Institute for Marine Science, Kiel.

5

\\\\\\\

Feed Types and Uses

INTRODUCTION

The high unit costs of aquaculture feeds reflect the carnivorous feeding habits of the majority of intensively farmed species of fish and shrimp. These animals are unable to utilize significant proportions of carbohydrates in their diets, which in most types of other livestock feeds provide cheap and available sources of energy (Table 5-1). As a consequence many feeds used in aquaculture are heavily dependent on the use of expensive marine products such as fish meals and oils to provide sufficient quantities of essential nutrients and supplies of energy. In addition, our knowledge of nutritional requirements of most farmed species is far from complete, a factor that leads to the excess use of certain essential ingredients in order to ensure that deficiencies do not occur.

Feeding animals in water poses problems in ensuring that essential nutrients are not lost during the feeding process. Aquaculture feeds are processed using special technologies to ensure that diets remain intact in water, before ingestion, and that soluble nutrients are prevented from dissolving. These factors also add to the manufacturer's costs.

Since animals farmed in intensive systems have little or no access to natural feeds, feed manufacturers must formulate feeds that are as nutritionally complete as knowledge permits, and still are cost effective. As the production of fish and shrimp through

Table 5-1   Comparison of Raw Material Sources of Pig and Poultry Feeds with Salmon and Shrimp Feeds

	Pig Grower %	Broiler Finisher %	Salmon Grower %	Shrimp Grower %
Carbohydrate sources	59.27	71.34	9.67	22.5
Plant proteins	31.00	10.00	—	20.00
Animal proteins	8.00	13.00	71.00	52.50
Synthetic amino acids	0.28	0.26	0.23	—
Fats	0.50	0.50	12.50	1.00
Minerals	0.45	0.60	—	1.00
Vit./Min. premixes	0.20	0.30	1.50	1.00
Other	—	—	—	2.00

From Evans, 1992

intensive methods has expanded here has been a parallel increase in nutritional research and improvements in food manufacturing technology.

Feed types used in intensive and semi-intensive aquaculture vary in terms of their formulation and methods of manufacture. Three general categories of feed are used.

WET FEEDS

Wet feeds comprise fresh or frozen, whole, chopped or minced trash fish. Types of fish commonly used as wet feed include herring, caplin, mackerel, blue whiting, and sand lance (Fig. 5.1). Less commonly wet feeds may be prepared from squid or other marine animals. In recent years the use of wet feeds has declined with the development and widespread availability of formulated feeds. Wet feeds are still used, primarily to feed certain marine species which respond well to wet fish diets, and for other species for which there is limited information available on which to base feed formulation. Sectors of the industry using some wet feeds include yellowtail and seabream culture in Japan, Atlantic cod farming in Canada and Norway, and grouper culture in Hong Kong and Singapore.

Although the use of wet feeds continues, their widespread use is limited by a number of factors. While the initial purchase cost of industrial fish may be very low, storage costs for freezing, and labor costs in handling, significantly inflate the real costs. In addition problems of nutritional balance may arise in feeding certain wet feed types to farmed fish. Most species used as feed are pelagic oily fish of variable quality, depending on the season in which they were caught, and on how they were handled and stored after capture. Oil content may be too high for some farmed species. For example cod develop enlarged, fat-filled livers if fed on a prolonged diet of mackerel. Some species used

Fig. 5.1 Wet feed. Chopped herring prepared for feeding to Atlantic cod, *Gadus morhua*, in Newfoundland, Canada.

as wet feed, such as herring and mackerel, contain high levels of the enzyme thiaminase, which if not destroyed by heat treatment, can lead to deficiency in levels of the vitamin thiamine, if these fish are fed over prolonged periods to growing fish. Disease organisms may also be transmitted through the use of unpasteurized wet feeds.

A commonly cited problem associated with the use of wet feeds is that of pollution. Clouds of fine particulate material are released from chopped or minced wet feed when thrown into the water. Wastage may also occur because wet feeds comprise a range of particle sizes; pieces may be too large or too small to be readily ingested by the fish; leading not only to wastage but also to potential deterioration of water quality. For this reason, the use of wet feeds is regulated against in some countries, particularly their use on freshwater fish farms. In Denmark, where farmers pioneered the intensive commercial production of rainbow trout, the industry was initially based on the use of wet feeds. Problems associated with pollution from fish farms contaminating rivers led to the introduction of regulations forbidding their use. This resulted in a progressive switch to the use of formulated dry feeds on Denmark's numerous trout farms. The replacement of wet diets is occurring in other sectors of the industry and can be expected to continue. Both moist and dry diets are being introduced into yellowtail culture in Japan (Doi, 1991), and formulated dry feeds are now commercially manufactured for Atlantic cod and other marine fish species.

Table 5-2   Moist Feed Formulation for Atlantic Salmon

Ingredients	Amount %
Ground pasteurized fish:	
Herring, caplin	44.0
Fish meal	22.0
Soybean meal	12.0
Wheat middlings	14.1
Choline chloride	0.4
Vitamin premix	1.0
Mineral premix	0.5
Herring, caplin or salmon oil	6.0

After Lall, 1991

WET AND MOIST FORMULATED FEEDS

Some of the problems encountered in using raw fish as aquaculture feeds are eliminated or reduced by using formulated wet and moist feeds. These enable fish farmers to take advantage of regionally available raw materials, which are mixed with other ingredients to give balanced feeds. Wet-pelleted feeds generally contain 50–70% moisture, and moist feeds 20–40% moisture. Both types of feed are based on ground, pasteurized, or ensilaged fish, mixed with binder meals and supplemented with vitamins, minerals, and oil. The first widely used moist feed was the Oregon Moist Pellet (OMP), developed in North America for use in salmonid hatcheries. They are generally cold-pelleted and must be used within several hours of preparation or frozen until required. A typical formulation used in the seawater cage culture of Atlantic salmon is shown in Table 5-2.

For some species of fish, moist feeds may be more palatable than dry feeds (Fig. 5.2). In cold-water aquaculture there is some evidence that moist feeds are more palatable at low temperatures than dry feeds, and some farmers find a better acceptance of moist starter feeds than dry feeds. The introduction of automated moist feed making equipment has also made possible the on-site manufacture of moist feeds using fish meal and a similar range of ingredients as are used in the commercial manufacture of dry feeds (Figs. 5.3, 5.4, and 5.5). Moist starter feeds are widely used in Pacific salmon hatcheries. They can also be manufactured on a small scale by farmers using relatively simple equipment.

Fish Silage

Fish silage is an alternative ingredient for minced fish for use in moist feeds. Fish silage is a liquefied product manufactured by chopping whole fish, or processing waste, and mixing with an acid. Chopping releases the natural enzymes present in the diges-

Fig. 5.2 Feeding eels. European eels, *Anguilla anguilla,* feeding from a wet paste on surface trays.

Fig. 5.3   Ingredient mixing. A large sack (1 tonne) of fish meal being blended with other ingredients in the preparation of a moist feed for Atlantic salmon in the Shetland Islands. Photo courtesy of Gerry Donovan

tive tracts of the fish, which rapidly digest and liquefy the tissues. The low pH prevents bacterial activity, and the resulting liquid silage is a relatively stable commodity that can be stored at ambient temperatures. This avoids the freezing costs associated with storing whole trash fish.

In untreated silage the enzymes continue to break down or hydrolyze the protein structures in the slurry. Over a period of weeks or months the original proteins are gradually reduced to a mixture of peptides and free amino acids, which reduces the nutritional value of the silage. Heat treatment (to 85°C) is necessary, following the initial stages of liquefaction to prevent this further breakdown. This inhibits enzyme activity and stabilizes the mixture. A number of acids are used in the preparation of fish silage.

Fig. 5.4 Moist feed preparation. An automated moist feed pelleting machine, linked to a hydraulic feed distribution system. Photo courtesy of Gerry Donovan

Fig. 5.5   Moist feed pellets. Farm-made, moist pellets being hydraulically conveyed from a mixer to offshore cages. Photo courtesy of Gerry Donovan

Weak organic acids (pH 3.5–4.5), such as formic or proprionic acids, have been shown to be most effective in preventing bacterial decay and fungal contamination.

DRY FEEDS

Dry feeds are by far the most commonly used feeds in intensive aquaculture. This largely results from convenience in their distribution, storage and handling, and consistence in quality. Feed mills producing dry aquaculture feeds are generally located close to the industry, and near ports for the importation of raw materials (Fig. 5.6).

Most dry feeds have less than 10% moisture content, which largely precludes bacterial activity. This gives dry feeds superior transport and storage characteristics (Chapter 6) when compared with wet or moist feeds. Current methods of manufacture eliminate most of the problems associated with antinutritional factors and also result in pellets which are relatively water stable.

There are two methods used commercially for the manufacture of dry feeds; steam pelleting and extrusion pelleting. While both methods are widely used, the more recently introduced extrusion or expansion pelleting methods result in feeds which are superior for some applications. Steam pelleting is however a cheaper process, and one for which most feed mills are equipped.

Fig. 5.6 Feed manufacture. A mill, purpose-built for the manufacture of fish feeds in Scotland. Photo courtesy of John Springate

Steam Pelleting

Finely ground ingredients are thoroughly mixed and treated with steam. Steam addition increases the moisture content to between 15–20% and the temperature to approximately 70–80°C. Under these conditions starch is partially gelatinized and functions as a binding agent. The mixture is then forced, under pressure, through a pellet die. Pellets are cut to length as they emerge through the die and are rapidly air dried (Fig. 5.7).

Extrusion Pelleting

During extrusion pelleting the ground, blended ingredients are processed at greater temperatures, moisture levels, and pressures than are used in conventional steam pelleting. The ingredients are first treated with steam. This brings the moisture level up to 20–30% and the temperature to 65–95°C. The mixture is then conveyed into a pressurized extrusion barrel (extruder) where it is cooked to a temperature of 130–180°C for 10–60 seconds (Fig. 5.8). Under these conditions starch present in the mixture readily gelatinizes. The cooked mixture is then extruded through a die plate. When the mixture emerges from the pressurized chamber, some of the water in the superheated mixture vaporizes, causing a rapid expansion in the volume of the pellets. The pellets are then cooled and dried. Some dietary ingredients, such as heat sensitive vitamin C

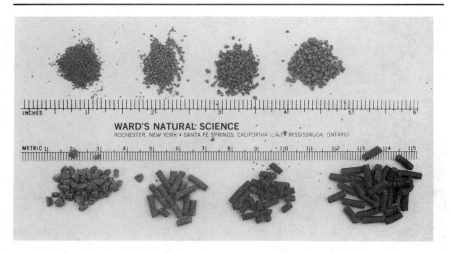

Fig. 5.7   Steam pelleted feeds. A range of pelleted feeds manufactured for shrimp.

and additional oils, phospholipids, pigments, and antioxidants may then be sprayed onto the absorbent surface of the pellets. The gelatinization of starch during extrusion, results in a more effectively bound, and hence more water stable pellet, than can be manufactured by steam pelleting. In addition, by controlling starch content, and conditions within the extruder, pellets with different densities can be made. Hence pellets which float, or which sink at different speeds, can be manufactured to meet various applications in aquaculture (extruded and steam-pelleted feeds are illustrated in Fig. 5.9).

A wide range of feed types is available for a some species, notably the salmonids, while more restricted ranges are available for most other fish and shrimp species. While formulations used by commercial feed manufacturers are often confidential and protected, so-called "open formulations" are widely available from public research organizations of universities and governments, and from international agencies which support the development of aquaculture. Examples of published formulations of dry diets for various species of fish and shrimp are shown in Table 5-3.

A current listing of feed types, taken from the sales list of a leading manufacturer of trout and salmon feeds, serves to illustrate the range of options available (Table 5-4). In established sectors of the aquaculture industry, for example, trout and salmon farming in Europe and North America, the differences in the quality of feeds from competing manufacturers may be small. They generally reflect some variation in the use of minor ingredients, the quality of the major ingredients, and in the overall energy balance of the feeds. In emerging sectors of the industry there may, however, be very significant differences in the quality of feeds from different sources.

Fig. 5.8 Extrusion equipment. A pressurized chamber used for the manufacture of **extruded** feeds.

Fig. 5.9 Dry feeds. An extruded salmon grower diet (left) compared with a steam pelleted diet (right). Note the more homogeneous appearance of the extruded feed.

COMMERCIAL FEED TYPES

Starter Feeds

Both live and formulated feeds are used in the hatchery production of juvenile fish and shrimp. The type of diets used for first feeding generally reflect the size and stage of development of the larvae after the contents of the yolk sac have been absorbed and the animals commence exogenous feeding. Most marine fish and shrimp larvae begin feeding soon after hatching when they are extremely small and may have incompletely developed digestive systems. The use of fine, particulate formulated feeds is limited for most of these species, particularly those which develop through a series of planktonic stages before metamorphosis into true juvenile stages. A succession of live food organisms, typically comprising diatoms and flagellates, 2–20 microns, followed by rotifers (*Brachionus plicatilis*), 50–200 microns, followed by brine shrimp nauplii (*Artemia sp.*), 200–500 microns, is obligatory for many species including shrimp and most marine finfish. Apart from a general HUFA requirement relatively little is known of the detailed nutritional requirements of the larvae of most marine species. It is common practice to nutritionally upgrade rotifers and brine shrimp by feeding them in suspensions rich in vitamins, protein, and essential fatty acids.

Table 5-3 Examples of Dry Feed Formulations Used in Fish and Shrimp Culture

Channel Catfish: Grower Diet	
Ingredients	Amount %
Fish meal (menhaden)	4.0
Meat and bone meal	4.0
Soybean meal	48.2
Grain or grain by-products	41.0
Dicalcium-phosphate	1.0
Fat	1.5
Mineral mix	0.05
Vitamin mix	0.25

From Lovell, 1992

Atlantic Salmon: Fry and Fingerling Feeds		
Ingredients	Amount %	
	Fry	Fingerlings
Fish meal (low temperature)	65.0	61.4
Blood meal	2.0	1.9
Skimmed milk powder	1.0	1.0
Whey powder	1.0	1.0
Grass meal	1.0	1.0
Seaweed meal	0.5	0.5
Soy lecithin	1.0	1.0
Extruded wheat	13.5	12.8
Fish oil (salmon)	11.0	16.5
Ground limestone	0.5	0.5
Salt	0.5	0.5
Vitamin and mineral premix	2.0	1.9

From Storebakken and Austreng, 1987

Shrimp *(Penaeus monodon):* Grower Diet	
Ingredients	Amount %
Shrimp meal	15.0
Fish meal	30.0
Soybean meal	15.0
Rice bran	15.0
Bread flour	15.0
Soya palm starch or corn starch	5.0
Oil (fish liver and soybean oil, 1:1)	4.0
Vitamin/mineral mix	0.95
Vitamin C	0.05

From New, 1990

Formulated Microdiets

The production of live food organisms for feeding fish and shrimp larvae is a major cost in hatchery operations; consequently, formulated microdiets have been extensively researched as potential replacements. Microdiets can be classified into three groups (Watanabe and Kiron, 1994):

- microencapsulated diets,
- microbound diets, and
- microcoated diets.

Microencapsulated diets contain ingredients in solution, colloid or in suspension surrounded by a digestible capsule, microbound diets are fine nutrient particles held together by a carbohydrate or protein binder, and in microcoated diets the nutrient particles are coated with digestible materials which render them impervious to water. Microdiets are commercially available in a range of sizes between 5 and 300 microns, and are now used in many hatcheries as partial replacements for live food organisms. Proprietary formulations are based on marine and animal proteins, plant proteins, mineral and vitamin premixes, pigments, antioxidants, and biodegradable binders or capsules (Watanabe and Kiron, 1994).

The first-feeding requirements of some freshwater fish species are less demanding. These include the salmonids, catfish and tilapias, which produce small numbers of relatively large eggs. These hatch to produce larvae which readily take fine particles of formulated food after their yolk sac is absorbed.

Fry and Fingerling Feeds

Dry feeds for fry and fingerling fish, and postlarval shrimp, are supplied in powdered form, as microcrumbles, crumbles, and small pellets. Diets formulated for the early rearing stages generally contain the same range of ingredients as diets for later stages but differ in their content of protein. The percentage of protein is generally highest for the first-feeding stages of fish and shrimp, and is then reduced for later stages. This supports the rapid gains in body weight seen in the early growing stages. In salmonids diets used for first-feeding typically contain 50–55% protein, diets for fingerlings contain 45–50% protein, and diets used in the main production phase contain 40–45% protein. Similar variations are used in commercial shrimp feeds (Table 5-5)

The main problems associated with the manufacture of starter feeds stem from the need to blend the full range of essential ingredients within a small particle. Conventional methods of manufacture involve crumbling larger pellets between corrugated rollers, and then screening and separating particles into different size ranges. More recently introduced methods of manufacture involve an agglomeration technique using

Table 5-4 Commercial Feed Types

Type	Variable
Fry and fingerling feeds	Crumb/pellet size/protein content
Floating or sinking	Pellet density
High energy	Protein/oil content
Low pollution	Ingredient digestibility
Pigmented	Carotenoid content
Broodstock	Protein/oil/vitamin/carotenoid content
Medicated	Antibiotic content
Winter feeds	Reduced protein

finely ground ingredients, which are spun in the presence of water to produce spherical particles. This process eliminates the need to grind larger particles, which reduces the dust and fines content, and produces particles of regular shape and of more uniform size (Figs. 5.10 and 5.11).

Floating Feeds

Pellets of low density, which float in water, are manufactured by adjusting the levels of certain ingredients, and controlling the physical conditions during extrusion processing. Pellets that float or sink at different speeds can be produced for various applications. Floating pellets are useful in estimating the feeding responses of fish grown in ponds where they cannot otherwise be readily observed during feeding. Extruded pellets were first used by catfish farmers in the United States where they are now widely used to bring fish to the surface during feeding. Direct observation of the feeding response enables the farmer to feed close to maximum intake levels while reducing the risk of overfeeding. The advantages of using floating pellets must offset their higher costs to the farmer. For pond aquaculture extruded floating pellets are more water stable than conventional pellets, and remain available for consumption over longer periods after feeding than do compressed pellets, which may sink into soft mud on the bottom of production ponds and be lost to the fish.

Table 5-5 Recommended Protein
Levels in Commercial Shrimp Feeds

Size of Shrimp (g)	Protein Level (%)
0.0– 3.0	40
3.0–15.0	38
15.0–40.0	36

Akiyama D. (personal communication)

Fig. 5.10 Fry feeds. An agglomerated feed (left) and a crumbled feed (right).

High-Energy Feeds

High-energy feeds have been developed primarily for use in salmonid culture and reflect the changes in formulation of feeds for these species which have been made in recent years. High-energy feeds contain high levels of lipids, 15–30%, which increase the total energy content of the diet and spare dietary protein for growth (Fig. 5.12). The trend toward high-energy diets for salmonids is seen in Table 1.3. High protein and lipid contents of these diets are reflected in higher costs. For any particular operation the farmer has to determine whether any improved performance in growth and food conversion offsets the increased cost (Chapter 10). Problems with high levels of visceral and muscular fat may arise when high-energy feeds are fed at maximum ration. This can result in a reduced quality score. It is often necessary for producers to starve fish prior to sale to reduce fat levels. Starvation periods may extend from several days to several weeks depending on water temperatures.

Low Pollution Feeds

Specially formulated feeds which are claimed to result in reduced pollution are produced by some manufacturers of dry feeds. The premise of reduced pollution is based on the use of high quality raw materials, which are more completely digested and result in reduced fecal wastes. Low pollution feeds are generally manufactured using the extrusion process, which results in more water stable pellets, and increases the digestibility of complex carbohydrates used to bind the feed.

Fig. 5.11 Bagging starter feeds. Photo courtesy of John Springate.

Fig. 5.12   High-energy feeds. A range of extruded pellets for use in the freshwater culture of salmonids.

Diets containing phosphorus compounds with high digestibility and low solubility are also being developed which meet the essential dietary requirements of fish for phosphorus while reducing the excretion of excess, or unavailable phosphorus. These developments are essential to reduce the environmental impacts of fish farming and to meet the regulations which in many countries control the levels of effluents, including phosphorus, that arise from fish farming operations.

Medicated Feeds

Feeds containing antibiotics are used to treat bacterial diseases of fish and shrimp. Their widespread use has raised concerns about the effects of antibiotics within the aquatic environment. The compounds which are used in aquaculture are similar to those used in human medicine, and the possibilities exist that pathogens transmitted in water or by consuming seafood may develop drug resistance if exposed to antibacterial drugs in fish farm effluents. Since the resistance to antibiotics can be transmitted from one bacterium to another there is a risk of transference of antibacterial drug resistance to normal bacteria in the human gut if sufficient numbers are ingested (Pillay, 1992). For these and other reasons the use of antibiotics is rigorously controlled in most countries, and efforts are made to minimize their application. For example, in the United States, Canada, and member countries of the European Community, approved drugs are available only on prescription by a licensed veterinarian, and must be administered under qualified supervision.

Fish are generally medicated by mixing antibacterial drugs with their food. Feed companies may make up batches of medicated feed under prescription. The minimum quantities that can be processed in a single commercial batch may, however, exceed a farmer's needs, and it is common practice for farmers to add antibacterials to their own food as required. This is normally accomplished by top-dressing feed with oil or 5% gelatin to which the medication sticks during mixing (Scott, 1993). Problems may arise in feeding measured quantities of medicated feeds to diseased fish, since appetite is often depressed, and some antibiotics reduce the palatability of treated foods for fish. One of the most widely used antibiotics, oxytetracycline, has been shown to reduce feed intake in healthy rainbow trout by 60% when fed at a level of 1% in a moist feed (Hustvedt et al., 1991).

The quantity of antibiotic to be administered is calculated on the basis of the daily feeding rate of the stock of diseased fish or shrimp. From a knowledge of the feeding rate and the total weight of the fish to be treated, the recommended amount of drug to be mixed with each kilogram of food can be calculated. Levels are generally adjusted so that fish receive approximately 1% of their body weight of medicated food per day, and additional feed amounts are made up with unmedicated feed. The duration of treatments normally ranges between 5–14 days (Scott, 1993). After treatment there is a statutory withdrawal period for each type of drug during which any residues clear from the tissues of the fish. Fish cannot legally be sold until the withdrawal period is complete and the fish are tested and approved for human consumption. The presence of antibiotic residues, detected in farmed fish or shrimp, may lead to the rejection of assignments by importing countries.

The range of antibiotics available for use is carefully restricted in many countries, and farm managers should seek early veterinary advice in the event of any disease outbreak where the causal agent is suspected to be a bacterium, and the use of medicated feeds may be predicated. The treatment of diseases in shrimp generally follow the same principles as those applied to fish. While a wide range of chemotherapeutants are used in the hatchery stages of shrimp production, medicated feeds containing antibacterials are primarily used for the treatment of diseases, such as *Vibriosis*, in juvenile and adult penaeids in nursery and growout ponds. In shrimp culture, the costs associated with manufacturing the necessary medicated feeds, and with delaying the harvest during withdrawal periods, often cannot be justified within the short production cycles typical of shrimp aquaculture. Under these circumstances farmers frequently seek to improve rearing conditions in an attempt to reduce the impact of a disease outbreak. This generally involves increasing the water exchange rates, using aeration devices, or partially harvesting the animals to reduce stocking densities.

General management issues involving the use of medicated feeds in aquaculture have recently been reviewed by Scott (1993) who has summarized a number of principles:

- Use antibacterials only when there is a strong likelihood of a bacterial infection.
- Start treatment as early as possible, having taken samples for bacterial examination.
- Use an antibiotic in as narrow a spectrum as possible in order to conserve normal bacterial ecology.
- Avoid prophylactic antibacterial therapy.
- Give the correct dose for the correct length of time to ensure proper tissue levels.
- Adopt a policy of restriction or rotation in the type of antibacterial agent used.

In practice the availability of diagnostic services, veterinary assistance, the legal status of drugs, and the overall cost factors associated with treating stocks of fish or shrimp, will dictate the pattern of use and effectiveness of medicated feeds.

The development and use of vaccines has significantly reduced the use of medicated feeds for the treatment of some bacterial diseases in trout and salmon culture. Feed additives such as glucan, a nonspecific immunostimulant, have also been shown to increase the general resistance of some fish species to disease. While effective vaccines are only available for a small number of diseases at present, it can be anticipated that new vaccines and immunostimulants will be developed in the future which will further reduce dependence on antibacterial drugs.

Pigmented Feeds

Feeds containing carotenoid pigments are used in the production of farmed salmonids to produce the pink-red flesh coloration, characteristic of wild-caught salmonids. Coloration is a significant factor in the marketability of these fishes, and with the exception of portion-sized rainbow trout produced in some countries, notably in the United States, the use of pigmented feeds has become a universally accepted practice in salmonid culture.

In addition to their use in feeds for salmonids, carotenoid pigments have also been evaluated for use in intensive shrimp culture to correct shell coloration (Menasveta et al., 1993), and in fish broodstock diets. Zeanthin pigments are supplemented in formulated feeds for farmed ayu in order to produce the lighter colored skin which is characteristic of wild fish (Kanazawa, 1991).

Only plants are able to synthesize carotenoids, hence fish and shrimp must obtain carotenoids from dietary sources. Pigments accumulate in the food chain, and wild salmonids obtain pigments through their consumption of zooplankton, such as *Euphausiids* (krill), which are rich in pigments derived from algae. Carotenoids comprise a widespread group of pigments; more than 500 chemical forms have been identified in nature. Of these, astaxanthin and its esters are the principal carotenoid pigments found in the tissues of wild salmonids, and in the exoskeletons of the decapod crustaceans (Torrissen et al., 1989).

Naturally occurring pigments in shrimp processing wastes were used initially to pigment farmed salmon. Results were often inconsistent, however, and involved feeding relatively large amounts of shrimp meal, or shrimp shell waste. The astaxanthin contents of shrimp meals vary, and they contain significant proportions of indigestible materials which lower food conversion rates, and increase effluent levels.

In the 1960s feed manufacturers turned to synthetic analogues of the natural pigments. Synthetic canthaxanthin, a pigment used elsewhere in the food industry, was shown to be readily absorbed and retained by fish, and produced coloration similar to that resulting from the deposition of astaxanthin. More recently a synthetic astaxanthin has been introduced which is a more effective pigmenter, and has become the pigment of choice of most feed manufacturers. Synthetic carotenoids are available as stable beadlets containing either 8% or 10% of pigment within a matrix of gelatin and carbohydrates, and protected by antioxidants. In addition to the availability of synthetic pigments, some species of yeast and algae containing astaxanthin have been evaluated as feed ingredients. Recently, strains of the yeast *Phaffia*, which have been selected for their carotenoid content, have become available commercially as ingredients for aquaculture feeds.

The use of synthetic carotenoids in fish feeds is regulated in most countries. In the member countries of the European Economic Community, feed manufacturers are permitted to use levels of 100 mg/kg of either astaxanthin or canthaxanthin in completed feeds, or a mixture of the two not exceeding 100mg/kg (Directive 70/524/ EEC). Their use is permitted only in trout and salmon exceeding 6 months of age, and is subjected to periodic review.

While synthetic pigments are readily available as feed additives they are expensive, and uptake levels are poor, estimated between 5–10%. Synthetic pigments cost more than $2000 per kg, and are normally added to feeds at a level of 50 g per tonne; this means that in diets costing $1000 per tonne, $90–95 or 9–9.5% of the cost of pigmented feed is wasted (Smith, 1990).

Various strategies are used for pigmenting salmonids. Generally fingerlings have only a limited capacity for carotenoid deposition in their flesh. Salmonids weighing less than 100 g will absorb pigment but most will be deposited in the skin, eyes, or internal organs. Larger fish deposit carotenoids in their flesh prior to sexual maturity, after which pigments are redistributed from the flesh to skin and eggs. Wild Atlantic salmon contain more than 12 mg astaxanthin/kg flesh, while Pacific salmon and Arctic charr may contain up to 30 mg astaxanthin/kg flesh. In general, farmed salmonids containing 5–10 mg pigment/kg flesh are acceptable to consumers. Fish destined for processing into fillets, for freezing, or for specialty markets may require higher pigment levels. Rainbow trout destined for sale as portion-sized fish are generally fed pigmented feeds containing 35–50 mg pigment/kg dry diet, after they reach a weight of 100 g and until they are ready for market. Atlantic salmon and sea-grown rainbow trout are normally fed pigmented diets containing 40–50 mg carotenoid/kg dry diet during the final 12

Fig. 5.13 Supplementary feed. A powdered mixture of locally available feedstuffs, fish meal, cotton seed cake, and rice bran, used in the semi-intensive culture of Nile tilapia and African catfish in Kenya.

months of the growout phase in seawater. Some Atlantic salmon farmers choose to use pigmented feeds throughout the seawater phase to ensure that any early maturing fish are pigmented to acceptable levels, and hence are salable.

Broodstock Diets

Broodstock fish and shrimp should be maintained under as close to optimal conditions as possible, and fed unrestricted rations of feeds containing high-quality ingredients. The diets for many species are supplemented with fresh wet feeds in an attempt to ensure an optimal nutritient balance and a full range of essential ingredients.

Increasing attention is being paid to the specific nutritional requirements of broodstock fish and shrimp, particularly to the affects of these nutrients on fecundity, and egg and larval survival (Bromage, 1995). Research with marine fish species has focused on a range of micronutrients, including essential polyunsaturated fatty acids (PUFA), particularly the n-3 series, docosahexaenoic acid (DHA); 22-6(n-3), and eicosapentaenoic acid (EPA); 20-5(n-3) and their derivatives; vitamin C; vitamin E; the carotenoid astaxanthin; and certain trace elements (Watanabe and Kiron, 1994).

Commercially available broodstock diets for fish are generally supplemented with vitamin C, vitamin E, trace elements, and pigments, above the levels normally found in production diets. Maturation diets for shrimp are most commonly based on squid meal supplemented with other animal proteins, essential polyunsaturated fatty acids, sterols, carotenoids, vitamins and minerals, stabilizers and attractants. In addition to using special maturation diets many shrimp farmers also feed fresh materials such as clam and squid muscle as the animals approach maturity.

Supplementary Feeds

The supplementary feeds used in semi-intensive aquaculture range widely in type from kitchen wastes to nutritionally complete formulated feeds. They may be in the form of single ingredients, mixtures of powdered ingredients (Fig. 5.13), or compounded into dough balls or pellets (De Silva, 1993). Most are farm-made, although supplementary feeds for some species are manufactured commercially.

The basis of supplementary feeding in semi-intensive aquaculture is to provide energy and nutrients in addition to those available from natural food organisms, in order to optimize growth. Supplementary feed formulations and ration sizes should therefore be selected on the basis of the density and total biomass of the cultured fish or shrimp, and the fertility and consequent availability of natural food organisms within the pond (Tacon, 1993). Supplementary feed requirements will generally increase as the production cycle progresses and the standing crop of fish or shrimp increases. In order to maintain energy and nutrient balance, and to support optimal growth, it may be necessary to provide supplementary feeds of different formulation, and to adjust ration sizes, as the availability of natural food organisms fluctuates during each production cycle.

Since there is little practical information available to guide manufacturers, most commercial feeds sold for use in semi-intensive culture are formulated as nutritionally complete feeds (Tacon, 1993). This is particularly the case in semi-intensive shrimp culture where the use of complete feeds carries the risk of over-feeding, particularly in the early stages of the rearing cycle.

REFERENCES

BROMAGE, N.R. 1995. Broodstock management and seed quality—general considerations. In Broodstock Management and Egg and Larval Quality, eds. N.R. Bromage and R.J. Roberts, pp. 1–24. Blackwell Science, Oxford.

DE SILVA, S.S. 1993. Supplementary feeding in semi-intensive aquaculture systems. pp. 24–60. In M.B. New, A.G.J. Tacon and I. Csavas (eds.) Farm-made aquafeeds. Proceedings of the FAO/AADCP Regional Expert Consultation on Farm-Made Aquafeeds. December 14–18, 1992, FAO/AADCP, Bangkok.

DOI, M. 1991. Yellowtail culture in Japan—an industry at the cross-roads. Infofish International 4:42–46.

EVANS, M. 1992. Constraints to the development of aquaculture diets in Australia—a feed manufacturers perspective. In Proceeding of the Aquaculture Nutrition Workshop, Salamander Bay, April 15–17, 1991, eds. G.L. Allan and W. Dall, pp. 214–220. NSW Fisheries, Brackish Water Fish Culture Station, Salamander Bay, Australia.

HUSTVEDT, S.O., T. STOREBAKKEN, AND R. SALTE. 1991. Does oral administration of oxolinic acid or oxytetracycline affect feed intake of rainbow trout? Aquaculture **92**:109.

KANAZAWA, A. 1991. Ayu, *Plecoglossus altivelis*. In Handbook of Nutrient Requirements of Finfish, ed. R.P. Wilson, pp. 23–29. CRC Press, Boca Raton.

LALL, S.P. 1991. Nutritional value of fish silage in salmonid diets. Bulletin of the Aquaculture Association of Canada **91**(1):63–74.

LOVELL, R.T. 1992. Nutrition and feeding of channel catfish. In Proceedings of the Aquaculture Nutrition Workshop, Salamander Bay, April 15–17, 1991, eds. G.L. Allan and W. Dall, pp. 3–8. NSW Fisheries Brackish Water Fish Culture Station, Salamander Bay, Australia.

MENASVETA, P., W. WORAWATTANAMETEEKUL, T. LATSCHA, and J.S. CLARKE. 1993. Correction of black tiger prawn (*Penaeus monodon* Fabricius) colouration by astaxanthin. Aquaculture Engineering **12**:203–213.

NEW, M.B. 1990. Compound feedstuffs for shrimp culture. In Aquatech '90. Shrimp Farming Conference Proceedings, June 1990, pp. 79–123. Infofish, Kuala Lumpur, Malaysia.

PILLAY, T.V.R. 1992. Aquaculture and the Environment. Fishing News Book, Oxford.

SCOTT, P. 1993. Therapy in aquaculture. In Aquaculture for Veterinarians—Fish Husbandary and Medicine, ed. L. Brown, pp. 131–152. Pergamon Press, Oxford.

SMITH, P. 1990. Innovations in salmon and shrimp feed. Aquaculture International Congress Proceedings. pp. 121–126. Aquaculture International, Vancouver.

STOREBAKKEN, T. and E. AUSTRENG. 1987. Ration levels for salmonids. I. Growth survival, body composition and feed conversion in Atlantic salmon fry and fingerlings. Aquaculture **60**:189.

TACON, A.G.J. 1993. Feed formulation and on-farm feed management. p. 61–74. In M.B. New, A.G.J. Tacon and I. Csavas (eds.) Farm-made aquafeeds. Proceedings of the FAO/AADCP Regional Expert Consultation on Farm-Made Aquafeeds, December 14–18, 1992, FAO/AADCP, Bangkok.

TORRISSEN, O.J., R.W. HARDY and K.D. SHEARER, 1989. Pigmentation of salmonids—Carotenoid deposition and metabolism. CRC Critical Reviews in Aquatic Science **1**: 209–225.

WATANABE, T. and V. KIRON. 1994. Prospects in larval fish dietetics. Aquaculture **124**:223–251.

6

\\\\\\

Feed Handling and Storage

INTRODUCTION

Optimizing handling and storage procedures for fish and shrimp feeds on farms is an essential component of good management practice. Feeds are valuable commodities, and those of the highest quality can readily spoil and denature if stored under inadequate conditions or for too long a period. Feeds past their "shelf life," or incorrectly stored, may not merely be unappetizing to fish or lacking in essential nutrients, but may contain toxic or antinutrient factors. Feeding fish or shrimp with old or contaminated feeds may result in abnormal behavior, poor feeding response and growth, and general loss of condition.

The two major factors affecting the quality of stored dry feeds are temperature and moisture. For these reasons the storage of aquaculture feeds generally presents fewer problems in cool temperate climates than under humid tropical conditions. Theft, and loss to, or contamination by, insects, rodents, birds, or other animals must also be prevented.

The different feed types described in Chapter 5 have different storage requirements.

WET FEEDS

Wet feeds may be delivered in bulk either fresh or frozen to farm sites. Raw fish should be examined for freshness since deterioration of fish begins immediately upon harvest, and will vary in extent depending upon icing or refrigerated storage conditions. Various changes can occur in fish prior to freezing. Degradation of protein by autolysis and glycogen through glycolysis may occur. In warm climates bacterial contamination will proceed rapidly if fish are not iced immediately on harvest. Fish are generally examined by visual inspection and by smell. Few farms have facilities for chemical examination, although rancidity tests may be appropriate if there are doubts, or if the fish are delivered to the farm pre-frozen. Such tests are used to quantify the degree of oxidation of fatty acids which has occurred in the fish.

Fresh raw fish should be frozen rapidly, in a plate freezer, and stored at low temperature. A process of rapid freezing reduces ice crystal development and prevents excessive rupturing of cells. Storage should ideally be at temperatures of −30° or lower. Such low temperatures will maintain the original quality of fish for 3−6 months. Common freezer temperatures of −18° are inadequate for long-term storage. While storage of fish in freezers will reduce enzyme activity, and inhibit the growth of bacteria, prolonged storage will result in some degree of fat oxidation and the denaturing of fat-soluble ingredients, such as certain vitamins and pigments. Exact storage conditions depend upon the type of fish being used. Lipid oxidation, the breakdown of fat in the presence of oxygen, will occur most rapidly in oily fish, such as caplin and mackerel. Hence these species cannot be stored for as long as less oily species, such as whiting. When thawed for use, either for direct feeding, or for mixing with other ingredients in a formulated food, fresh fish should be used as rapidly as possible before significant bacterial contamination can occur. This is generally within a few hours after thawing. To minimize the risk of bacterial contamination some farmers prefer to use wet feeds as a frozen block. The block is floated on the surface of the cage or tank, where it gradually melts at ambient water temperature, releasing feed over a prolonged period. This method is claimed to result in less wastage and pollution from uneaten food scraps.

MOIST FEEDS

Storage procedures for moist feeds, containing minced raw fish, or ensilaged fish, parallel those outlined above for wet feeds. Once mixed and pelletized, moist feeds should either be fed within several hours, or frozen. Commercially produced moist feeds, such as the Oregon Moist Pellet, used at some hatcheries in the culture of Pacific salmon, are delivered frozen and stored frozen until used. Freezer units of various types and sizes are used to hold moist feeds at temperatures of −20° or below. Most hatchery freezers are fabricated as part of the hatchery buildings with walk-in or drive-in capacity.

In some regions moist feeds are manufactured commercially and delivered locally, from the feed mill to the farm, on a daily basis, or prepared and used freshly on-site (Fig. 6.1). The use of pasteurized fish and oil antioxidants may extend the shelf life of moist feeds to 2−3 days if kept cool. Ideally, however, they should be used within 24 h after manufacture.

Fig. 6.1 Moist feed distribution. Hydraulic distribution pipes for moist feeds manufactured on-site at an Atlantic salmon farm in the Shetland Isles.

DRY FEEDS

Storage times for dry feeds depend upon a number of factors. These are related to formulation and manufacturing methods, and to general storage conditions. These factors probably account for the disparity in published recommendations, for example, Beveridge (1987) suggests storage times, for dry compounded feeds, of up to 9 months in temperate zones and 2–3 months in tropical zones, while New (1987) recommends 1–2 months storage for dry feeds stored in either tropical or temperate regions. In practice, the storage times for dry diets, after manufacture, depend on a number of factors. The main factors are diet type, the quality of fats used in their manufacture, and the nature and conditions within the farm's storage facilities.

Dry diets are most commonly delivered to farms in 25 kg bags or less commonly in 50 kg bags. Most feed manufacturers also offer bulk delivery of feed, for storage in hoppers. Bulk dry feeds may be delivered in 500 kg or 1 tonne plastic sacks (Fig. 6.2), or alternatively in plastic totes. These typically hold up to 500 kg and are returned to the food supplier where they are washed, sterilized, and reused. Feed is normally bagged in either plastic sacks or multilayered paper sacks with plastic liners. Plastic sacks are more expensive but have a number of advantages. In particular they are waterproof, enabling food to be stored temporarily outdoors. This may be useful at remote sites, such as offshore sea cages, where food sufficient for several days may be delivered in a single trip

Fig. 6.2 Bulk feed sacks. Dry fish feed packed in large plastic sacks (1 tonne capacity) for bulk delivery. Photo courtesy of John Springate.

from facilities on shore, and held in temporary storage without risk of getting wet. Formulated dry feeds, particularly those with high oil content, are softer than conventional livestock feeds and hence more prone to damage if poorly handled. Care must be taken not to drop or throw sacks of feed or to allow farm staff to walk across stored sacks. Sacks of feed are best delivered stacked on wooden pallets on which they can be stored and moved as necessary using forklift trucks or pallet jacks. Stacks of 25 kg feed bags should not exceed 8–10 in number in order to avoid crushing feed stacked at the bottom (Fig. 6.3).

Dry feeds delivered to the farm in bulk are stored in silos. These are equipped with outlet shoots suitable for loading either trucks or boats. There are small savings for the

Fig. 6.3 Dry feed packed on pallets. Delivering feed by boat on 1 tonne pallets (40 × 25 kg sacks) to a salmon farm in British Columbia, Canada.

farmer when purchasing feeds in bulk since packaging and handling costs are reduced. However, food stored in bulk silos or hoppers is more prone to moisture problems than food packed in sacks, and storage times should be reduced.

Storage Facilities

Dry feed should be stored under cool, dry conditions, ideally at temperatures below 20°C and at relative humidities below 75%. The building should be used exclusively for feed storage and planned to incorporate the following features:

- Secure against theft.
- Adequately ventilated and waterproof.
- Screened against rodents and birds.
- Of sufficient size to permit feeds to be stored in clearly marked batches, according to type and date of purchase.
- Conveniently situated for taking delivery of feeds and for their distribution on the farm.
- Isolated from the fish production areas, with separate access, to reduce the risk of disease transmission by delivery vehicles or personnel.

Construction of Food Stores

Buildings used for storing aquaculture feeds must be designed to prevent the exposure of feed to harmful levels of moisture or excessive temperatures. The insulating properties of the construction materials and the pattern of ventilation within the building are critical factors in regulating both temperature and moisture.

Insulation

In regions with extremes of heat or cold, attention must be given to storage temperatures. Buildings should be adequately insulated, and where necessary, heating or cooling systems incorporated. There are three mechanisms involved in thermal transmittance: conduction, convection, and radiation. Conduction is the passage of heat by particle to particle contact, convection involves heat flow by air mass movement within materials, and radiation is the passage of heat by waveform. To provide effective insulation, or heat barrier, each of the mechanisms must be minimized. This is done by the choice of materials with poor particle contact characteristics, many small trapped air pockets, and reflective surfaces of silver or white coloration. Insulation materials should also be waterproof or vaporproof, otherwise water will replace the air masses and significantly increase heat conductivity (Carpenter, 1988). While values of thermal transmittance can be expressed in various ways a commonly used measure, applied to building materials, is the "U" value. U values are expressed as heat passage (watts per square meter per degree Celsius)(W/°C m^2) through a surface area under the "pressure" of a given temperature gradient. They are used in the equation:

$$Q = UA \, (ti - to) \tag{6.1}$$

where **Q** is the quantity of heat passing, **U** is the structure thermal transmittance, **A** is the structure area, and **ti** and **to** are the temperatures on the inside and outside faces.

Examples of U values for commonly used building materials are given in Table 6-1. Higher values indicate poorer performance as insulators. U values are affected by the nature of the external aspect and environment and do not represent absolute values for a material. Exposure, color, cleanliness, and dampness will all affect U values (Carpenter,

Table 6-1 Examples of U Value on Thermal Transmittance

Wall	U W/°C m²	(R)
215mm dense, hollow, concrete blockwork	2.05	(0.49)
As (1) + outside rendering + 12mm expanded polystyrene slab	1.14	(0.88)
275mm "cavity wall" of 100mm dense concrete blocks + 25mm expanded polystyrene slab	0.68	(1.47)
215mm solid wall of foamed blockwork + rendering both sides	0.45	(2.22)
6mm cement-fiber sandwich on 50 × 50mm timber frame with expanded polystyrene core	0.45	(2.22)

Roof		
Corrugated cement-fiber sheet	6.53	(0.15)
Corrugated sheet steel + fiber "insulation" board	4.82	(0.21)
Corrugated cement-fiber sheet + 60mm mineral wool (+ vapor barrier) + 4.5mm cement-fiber lining board	0.60	(1.67)
Corrugated cement-fiber sheet + 55mm extruded, foil-faced, polystyrene sheet	0.59	(1.69)
Corrugated cement-fiber sheet + 40mm foil-faced polyurethane board	0.50	(2.0)

Floor	R_{f45} (°Cm²/W)
Dense concrete floor	0.042
Concrete slatted panel	0.086
18mm screed on 150mm lightweight aggregate	0.17
Wooden slats, 58 × 70mm with 10mm gaps	0.23

Note: R_{f45} is not directly equivalent to R (= I/U) above as it takes into account sideways heat movement from "point" body contacts.

NB: Insulation value is dependent on the moisture content (or lack of it), thus a water/vapor proofing agent is essential for a continuing satisfactory insulation level from a material.

From Carpenter, 1988

1988). The thermal conductivity or K value (W/m°C) is another measure of heat transmission through materials. These are absolute values and are quoted per unit of thickness.

In well-insulated buildings freezing temperatures can generally be avoided by the use of simple infra-red emitters (Fig. 6.4), while extreme high temperatures may necessitate control by the installation of air conditioning units.

Fig. 6.4   Feed storage. A fully insulated feed store, with food correctly stored on pallets in a single layer. Infra-red emitters are in use to prevent freezing temperatures.

Ventilation

Ventilation or air exchange should be provided in food stores to lower temperature and to control humidity. Temperature and air moisture content are linked by the fact that relative humidity is dependent on temperature levels. With rising temperature the amount of moisture in an air mass required to cause saturation increases. As temperature decreases so the saturation level decreases and with a given water mass, known as the specific or absolute humidity, the relative humidity will increase, until at 100% any further temperature reduction will cause condensation. This is referred to as the dew point and is important for insulation design with respect to vaporproofing (Carpenter, 1988).

Ventilation in feed stores can be provided by natural means, such as an open front or open ridge (Fig. 6.5); partly controlled, using exhaust fans; or totally controlled, using complex duct or pressurizing fan equipment. Natural ventilation is most commonly used with airflows resulting from the effects of wind or temperature gradients (stack effect). Exhaust fans may be necessary where natural ventilation is inadequate. These give more control but have limited value for partially open buildings with adverse temperature gradients. Pressurized methods of ventilating are most effective and

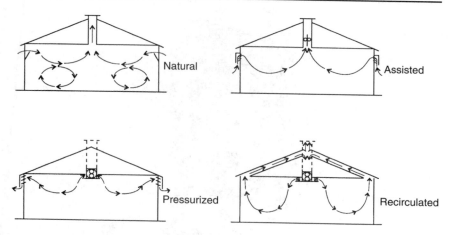

Fig. 6.5 Ventilation systems. After Carpenter, 1988.

offer the greatest measure of control. In these systems, air is either sucked or blown into the building. Such systems are the most costly, however, and require carefully designed duct systems to distribute the air evenly. Heat exchanged through ventilation may be shown as:

$$\text{Heat loss or gain (Q)} = m \times s \times (t_e - t_i) \qquad (6.2)$$

where **m** is the mass of air exchanged per unit time, **s** is the specific heat of air (0.36 J/Kg) and **te** and **ti** are the temperatures of exhaust and inlet.

Storage Hoppers and Silos

Bulk storage systems are widely used for storing dry aquaculture feeds, and for storing silages and binder meals used in the on-site manufacture of moist feeds. Silos used for storing dry feeds range in volume from 2–30 m³, and are typically prefabricated in galvanized steel. Silos used for storing silage are generally larger, up to 60 m³, and must be internally coated with acid-resistant material in order to prevent corrosion. Both types are available in single-unit construction, or in hoop or ring form, which can be assembled on the farm site (Figs. 6.6, 6.7).

The use of storage hoppers should enable the farmer to buy food in bulk at discounted prices (Chapter 10). Bulk food may be delivered loose, by road transport, or where this is not possible, in large plastic sacks. Silos are available with large diameter hatches to accommodate filling from bulk sacks. Where bulk feed is used in conjunction with automated or mobile feeding systems the time and costs, normally associated with handling 25 kg bags (Fig. 6.8), should be significantly reduced. Problems associa-

Fig. 6.6   Bulk storage. A feed storage hopper alongside channel catfish ponds in Louisiana. Photo courtesy of Jay Huner.

ted with dust and crumbs in bulk feeds can be reduced by the use of bucket-type elevators. These are more expensive than the normal pneumatic filling systems but reduce the physical damage to pellets. A disadvantage in the use of bulk storage is the reduced flexibility for storing a range of feed types or pellet sizes.

MAINTENANCE OF FEEDSTOCKS

Feed supplies should be purchased freshly and on a regular basis, to avoid the problems associated with long-term storage. Purchase and delivery intervals will depend on the regional availability of feeds and the logistics of delivery. Ideally feeds should be purchased and used on a monthly basis (Akiyama and Chwang, 1989), and batches of feed should be clearly identifiable to ensure that older stocks are used before new. Feed bags should be stored on pallets, off the ground, and away from walls to allow air to circulate and to maintain even temperatures. The walls and floor of feed stores should be

Fig. 6.7 Bulk storage. Dry feeds for Atlantic salmon stored in silos on a floating barge in Maine, USA.

kept clean and any spilled materials disposed of. Spilled medicated feed, containing antibiotics, should be disposed of in accordance with regional regulations or guidelines.

The temptation to store chemicals in the farm food store should be resisted, since any spills or accidents may contaminate feedstocks, which will be expensive, and possibly difficult to replace quickly.

Prior to use, feeds should be checked for any changes in texture, color, or smell (Fig. 6.9). Dry feeds should be free flowing. Any clumping of food may indicate moisture problems or fungal infection. Off-smells may indicate either fat rancidity or staleness. Feed should never be stored in direct sunlight since this may adversely affect the potency of vitamins and the quality of lipids (Fig. 6.10).

Feeds should be carefully checked for the presence of any mold or insects, or any rodent damage. If present the infested feed sacks should be isolated and the necessary steps taken to prevent further problems. Feces and urine from rats and mice may contaminate large quantities of food, greatly in excess of the feed quantities they may consume. Rats are persistent pests that are difficult to control or eliminate. Their bodies, feces, and urine carry diseases including leptospirosis, murine typhus, and salmonellosis (Marriot, 1985). Workers handling feeds should be made aware of the potential disease risks and instructed to cover any unprotected cuts or scratches. Spaces where rats can enter the storage building should be sealed off and the areas around the food store should be well-drained and free of debris or pockets where rats can hide and breed (Minor, 1983). Doors should fit tightly, and if not steel, should have a 20 cm metal

Fig. 6.8 Handling dry feeds. Farms should be planned to take into account the routine activities involved in feeding. The positioning of feed stores and provision of full access for feeding equipment should be planned to reduce feed-handling times.

strip on the bottom to prevent rats gnawing entrance holes. A single rat is capable of eating and spoiling hundreds of dollars worth of food each year. Rodents infesting food stores should be trapped rather than poisoned, eliminating any risk of further contamination of food supplies.

Where feeds are stored for too long a period, or under poor conditions, serious problems may arise from loss of vitamins, contamination with adventitious toxins, and from the effects of oil rancidity.

LOSS OF VITAMINS

Vitamins are biologically active compounds which are generally sensitive to their physical and chemical environment. A range of factors including temperature, pressure,

Fig. 6.9 Poor storage of shrimp feeds. Expensive feed exposed to light, high temperatures, and excessive humidity and temperatures. Food stored outside is also vulnerable to theft, rodents, and pest infestation.

friction, moisture, conditioning time, feed composition, and light can adversely affect vitamin stability during feed processing and storage (Coelho, 1991). In premixes, trace elements catalyze oxidation and reduction reactions which can lead to reduction or loss of vitamin activity, and during manufacture friction, temperature and pressures will variously affect vitamin stability. Vitamins are subject to less friction in an extruder than they are in a pellet mill, however high temperatures, pressure, and moisture during extrusion may cause significant vitamin loss. Table 6.2 lists average stability values for vitamins in premixes, pelleted, and extruded feeds.

In stored food, moisture and oxidation by polyunsaturated fatty acids, peroxides and trace minerals can affect the stability of vitamins. Vitamin supplementation rates in formulated feeds are adjusted by manufacturers to compensate the losses associated with manufacturing and limited storage periods. This does not however diminish the need to follow correct storage procedure.

ADVENTITIOUS TOXINS

Fungal or mycotoxins are the most commonly occurring feed contaminants. Adventitious toxins are produced by molds or fungi which may grow on feedstuffs used in formulated feeds. The most common fungi found in feeds are species of the genus *Asper-*

Fig. 6.10 Checking shrimp feeds. All stored feeds should be routinely checked before use for off-smells, mustiness, and any discoloration or clumping.

gillus. Various ingredients, including cottonseed meal, peanut meal, corn products, and cereal grains, are susceptible to *Aspergillus* infestation. Where used in animal feeds these ingredients are routinely tested for contamination with aflatoxin B[1] (AFB), an adventitious toxin produced as a biosynthetic product by *Aspergillus flavus* and *Aspergillus parasiticus*. The toxin is a powerful liver carcinogen which has been demonstrated to produce tumors in rainbow trout at levels as low as 0.006 ppm. Proprionic acid, benzoates, or formic acid may be added to diets as mold inhibitors. In the routine checking of feeds, care should be taken to check for signs of mold infestation.

The following conditions in stored feed are indicative of possible mold infection (Jauncey and Ross, 1982):

- Mustiness or staleness.
- Discoloration and lumpiness.
- An increase in moisture content and temperature, with resulting "sweating."

Feed showing any signs of mold should be discarded.

OIL RANCIDITY

Oxidative rancidity is generally considered to be one of the most serious detrimental changes which can occur in stored feeds. In the absence of suitable antioxidant protection, lipids rich in polyunsaturated fatty acids, including the essential fatty acids, are highly susceptible to auto-oxidation. During the process of auto-oxidation, a number of harmful breakdown products are produced, including free radicals, peroxides, hydroperoxides, aldehydes, and ketones. These compounds in turn react with other nutrients, including proteins, vitamins and other lipids, reducing their biological value and availability following digestion. The pathological effects of feeding oxidized oils to fish have recently been summarized by Tacon (1992). Symptoms include reduced growth, and feed efficiency, swollen or fatty livers, and increased mortality. These effects are generally preventable by supplementing feeds with vitamin E, a natural antioxidant, or synthetic antioxidants. Commercial feeds are usually supplemented with one, or a combination, of the synthetic antioxidants, butylated hydroxyanisole (BHA), butylated hydroxytoluene (BHT) or ethoxyquin. These are typically added at levels of 0.015–0.02% of the diet.

Feed Labeling and Quality Control

Products are licensed and feedstuffs are labeled in accordance with national regulations. Labeling requirements differ between countries but generally list the ingredients and stipulate the minimum and maximum levels of the various ingredient classes used in each particular food type. For example, shrimp feeds manufactured in the Philippines must be labeled under the information categories shown in Table 6.3. In most countries quality control procedures are in place to verify national standards established for both feed ingredients and finished feeds. Hence the information on feed labels provides some guarantee for purchasers of the freshness and quality of feeds. If farmers, for whatever reason, doubt the quality of purchased feeds they should take representative samples to be submitted for physical and chemical examination. This may be done by taking a scoop from every fifth bag of 40–50 bags and pooling the samples. The sampled feed should then be placed in tightly sealed containers and sent to an approved

Table 6-2 Vitamin Stability in Premixes and Pelleted and Extruded Feeds

	Stability				
Vitamin:	Very High	High	Moderate	Low	Very Low
	Choline Chloride	Riboflavin Niacin Pantothenic acid E Biotin B12	Thiamin mononitrate Folic acid Pyridoxine D3 A	Thiamin HCL	Menadione Ascorbic acid
			Losses/Month		
Premixes without choline and trace minerals	0%	<.5%	.5%	1%	1%
Premixes with choline	<.5%	1%	3%	7%	10%
Premixes with choline and trace minerals	<.5%	5%	8%	15%	30%
Pelleted feed	1%	3%	6%	10%	25%
Extruded feed	1%	6%	11%	17%	50%

From Coelho, 1991

Table 6-3 Labeling Regulations for Shrimp Feed (Bureau of Animal Industry, Department of Agriculture, Philippines)

Net weight
Name and address of maker
Brand or trademark

Proximate composition:
  Minimum % of crude protein and crude fat
  Maximum % of crude fiber, ash and moisture
  Maximum % of minerals (if more than 5%, the levels of
    Ca [Calcium] and Ph [Phosphorus] must be indicated)
  Guaranteed aflotoxin level
  Recommended stocking density
  Guaranteed peroxide level
  Registration no.

Additional for medicated feeds:
  Name and percent of drug
  Directions for use
  Safety warnings
  Withdrawal periods

Control no.
Date of manufacture
Expiry date
Specific warehousing conditions

Reproduced with permission

laboratory for testing. The feed supply company should be informed immediately in the event of any apparent problems with feed quality.

Feed Inventories

Complete records should be kept for each batch of feed delivered to the farm. These should always include details of date of delivery, manufacturer, feed type(s), batch numbers, quantity, cost, and any observations on the condition of the food on receipt.

REFERENCES

AKIYAMA, D.M. and N.L.M. CHWANG. 1989. Shrimp feed requirements and feed management. In Proceedings of the SE Asia Shrimp Farm Management Workshop, ed. D.M. Akiyama, pp. 75–82. American Sobean Association, Singapore.

BEVERIDGE, M.C.M. 1987. Cage Aquaculture. Fishing News Books Ltd., Farnham.

CARPENTER, J.L. 1988. Farm buildings. In The Agricultural Notebook 18th Ed., ed. R.J. Halley and R. J. Soffe. Butterworth Scientific, London.

COELHO, M.B. 1991. Effects of processing and storage on vitamin stability. Feed International **12**:39–45.

JAUNCEY, K. and B. ROSS. 1982. A Guide to Tilapia Feeds and Feeding. Institute of Aquaculture, Stirling.

MARRIOT, N.G., 1985. Principles of Food Sanitation. Van Nostrand Reinhold Company, New York.

MINOR, L.J., 1983. Sanitation, Safety and Environmental Standards, AVI Publishing Company, Inc. Westport.

NEW, M.B. 1987. Feed and feeding of fish and shrimp. ADCP/REP/8726 p. 275, FAO, Rome.

TACON, A.G.J. 1992. Nutritional fish pathology—Morphological signs of nutrient deficiency and toxicity in farmed fish, p. 75. FAO Fisheries Technical Paper 330. Rome, FAO.

7

⟍⟍⟍⟍⟍⟍

Feeding Methods

INTRODUCTION

It is not uncommon to find neighboring fish farms, culturing the same species in similar holding systems, but adopting widely different feeding methods. Since the primary objectives for most fish farmers are to produce high-quality fish at least cost, these observations indicate that the choices made on farms are based on a diverse range of factors and experiences. The widespread availability of durable dry feeds and rapid advances in control technology has led to the introduction of numerous types of automated systems for delivering feed to fish. Options for fish farmers range from traditional hand feeding methods to the use of automatic, computer-controlled systems. In practice, for reasons outlined below, farmers frequently choose to use a combination of feeding methods.

The first choice confronted by fish farmers is whether to feed fish by hand or whether to use labor-saving mechanical or automatic systems. This decision is based on a number of factors including:

- Labor costs,
- The scale of farm operation,
- The species being farmed,
- The type of holding system, i.e., ponds, cages, tanks, or raceways, and
- Hatchery or grow-out operation.

An initial choice may be made on straightforward economic factors, such as the cost of local labor. In the shrimp industry of Southeast Asia, animals are fed almost exclusively by hand, whether from the pond banks or from boats. The choice of feeding sys-

Fig. 7.1   An offshore submersible cage. Feeding from the storage hopper at the top of the structure is remotely controlled from a shore base. Photo courtesy of Laura Halfyard.

tem may alternatively be dictated by the physical nature or location of the holding systems used in fish culture. The dimensions and accessibility of large offshore cages, for example, necessitate the use of automated feeding systems, which can be controlled from a shorebase (Fig 7.1).

In examining the merits of the various feeding methods available to fish farmers, the factors discussed in Chapter 2, concerning the diverse feeding habits of the various species and the factors controlling their appetite and feeding behavior, should be taken into account. Various features of feeding systems should be appraised for any particular culture operation. These should include

- The spread, or effectiveness with which the food is distributed within the holding system.
- The feeding intensity, or the volume of food the equipment is capable of delivering in a single feeding event.
- Storage capacity.
- Cost.
- Reliability.

FEEDING METHODS

A general description of the various feeding methods is given below with a summary of their relative advantages and disadvantages. Feeding methods for shrimp are discussed under a separate heading.

Hand Feeding

Distributing feed by hand allows close observation of the appetite and feeding behavior of those fish species which feed actively at the surface, and are grown in clear-water systems. An experienced operator can scoop feed evenly over a limited surface area, in response to the apparent demands of the fish (Fig. 7.2). Problems associated with overfeeding can generally be avoided if feeding is gradually reduced as the fish approach satiation. Effective hand feeding necessitates some initial training and experience if food wastage is to be avoided. The main advantage of hand feeding is that when carried out carefully it is sensitive to the changing feed requirements of the fish. Appetite can change in response to a wide range of both physiological and environmental factors and significant daily fluctuations can occur. The practice of handfeeding also ensures that the behavior of the farmed fish is regularly observed and that any abnormalities can be addressed (Fig. 7.3). The main disadvantages of hand feeding are that it is labor intensive and time consuming and may be limited in its application on large farms.

Fig. 7.2 Hand feeding salmon. Broadcasting moist feed to broodstock Atlantic salmon on a farm in the Bay of Fundy, Canada.

Fig. 7.3   Hand feeding tilapia. Dry feeds being spread by hand to cage-reared tilapia in Java, Indonesia.

In practice most fish farmers retain some hand feeding as part of their feeding policy in order to maintain close and regular observations of their fish. These advantages of hand feeding may be combined with the use of mechanical or automatic feeding systems.

- Farm managers may, for example, feed 70–80% of estimated daily feed requirements using special equipment, and then complete feeding by hand in order to accommodate daily variations in the fishes' appetite.
- Alternatively a small number of holding units can be fed exclusively by hand and a careful record made of food consumed. This information can then be used in programming automatic feeders or in calculating feed amounts for loading demand feeders.

These policies can result in significant food savings and reduced wastage.

Mechanized Feeding

Mechanized feeding systems are widely used in aquaculture and many new types are being developed and tested (Singh, 1992). Most systems are designed to operate with dry feeds although some can handle moist feeds. Their capital and operating costs may be offset by reduction of the labor costs associated with feeding by hand, along with other potential advantages. With appropriate timing controls, automated systems can be set to deliver feeds at any time, at any given frequency, and in accurately determined quantities. Powered distribution systems can also spread feed more evenly, and over larger areas, than is possible when broadcasting feed by hand.

There are two basic types of mechanized feeding system:

- Demand feeders
- Automatic feeders

Demand Feeders

Demand feeders, also referred to as response feeders, most commonly comprise feed hoppers fitted with a pendulum and food release valve. The pendulum is attached to some type of valve mechanism that controls the flow of food from the hopper (Figs. 7.4 and 7.5). When fish bite or hit the pendulum a controlled quantity of food is released into the water. In theory, fish trained to operate the pendulum are able to access food on demand. A second type of demand feeder is the plate feeder, designed for feeding small fish unable to operate a pendulum. This type of feeder has a submerged feed tray linked to a hopper. When food is consumed from the tray, a trigger releases more food from the hopper. Again the feeder is sensitive to the feed demands of the fish.

Although demand feeders are widely used by fish farmers there have been relatively few controlled studies on the behavior patterns of fish with access to self-demand feeders. Demand feeders have been used in isolated studies to examine daily and seasonal feeding patterns (Adron et al., 1973; Grove et al., 1978; Boujard and Leatherland, 1992; Eriksson and Alanärä, 1990). The fishes' response to demand feeders is based on positive reinforcement of individual behavior. Results from studies conducted with rainbow trout indicate that experimental groups learn fairly rapidly to activate the pendulum and access food. Results from experiments with rainbow trout (Landless, 1976) and chinook salmon (Crampton et al., 1990) have shown that these species take between 7–10 days to reach a stable level of self-feeding.

Demand feeding systems are also available which link a touch sensitive probe through a control system to an automated feed dispenser. When activated by fish the feeder dispenses a predetermined amount of food from a hopper by a vibrator or a food spreader. The advantage of these systems is that they offer better food distribution than conventional demand feeders, which drop feed onto the water surface immediately below the feeder. The response level of the sensor and the running times of the feeder are adjustable to accommodate use with different species or sizes of fish. These systems combine some of the advantages of demand feeders with those of automatic feeders (Fig. 7.6).

The potential advantages of using demand feeders are first, that they should make food continuously available, and second, that the fish should be able to consume a meal of appropriate size to satisfy their appetite. Demand feeders are the only feeding system

Fig. 7.4   Demand feeders in use at a rainbow trout farm in Hampshire, England.

with this potential and as such merit further study and development. To work effectively, demand feeders must be readily accessible to the fish. This means that careful attention should be paid to the numbers of feeders operating in each tank or pond. It is common to see a single demand feeder operating in each fish holding unit, which may be inadequate to provide free access to fish seeking food. Conventional demand feeders have limited application at exposed cage sites where waves and water currents may trigger the pendulum, releasing excess feed. Sensor-operated demand feeders, however, are insensitive to wave and wind action.

Basic demand feeders are relatively cheap to buy or simple to construct. A wide range of types are commercially available with hopper sizes ranging typically from 15 –

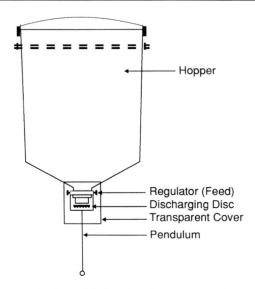

Fig. 7.5   An illustration of a demand feeder. From Senn et al., 1984.

75 liters capacity. Most have an adjustable mechanism that controls the volume of food released at each activation of the pendulum.

Studies on growth and feed conversion of trout feed using sensor-operated demand feeders in cage systems have shown that fish given unrestricted rations displayed higher growth than fish given restricted rations, but at the expense of a higher food conversion ratio (Alanärä, 1992a,b). In the same series of experiments a comparison was made of timer-controlled automated feeders and demand feeders delivering restricted rations. The results showed that rainbow trout (1.1–1.2 kg) given small repeat feeds at short intervals (4–5 min) from automatic feeders displayed poorer growth and higher food conversion ratios than fish fed similar rations from demand feeders. These results probably reflect the greater energy expenditure of fish constantly seeking food and rarely reaching satiation.

Automatic Feeders

Automatic feeders dispense predetermined amounts of feed at set intervals and vary widely in the sophistication of their design and operation. Most are electrically powered, from batteries, solar cells, or mains supply, and are operated from a central control unit. Food may be spread as it falls from the hopper by revolving or vibrating mechanisms (Fig. 7.7) or be blown by compressed air (Figs. 7.8 and 7.9). Depending on the mechanism, food may be spread in a variety of patterns; most commonly in a circular or semicircular pattern, or in a pie-shaped area out from the feeder. Feeders may be

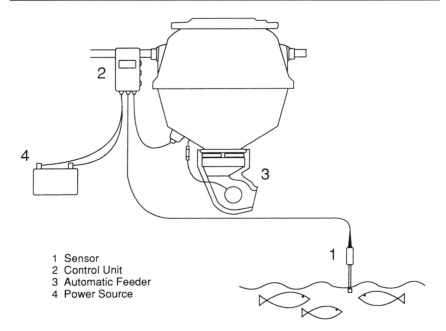

1 Sensor
2 Control Unit
3 Automatic Feeder
4 Power Source

Fig. 7.6 A demand-feeding system. A sensor is linked to an automatic feeder which spreads feed more effectively than mechanical demand feeders. From Eriksson and Alanärä, 1990.

positioned centrally or at the side of the fish holding unit. Many automatic feeders incorporate photocells in their control systems to ensure operation during the full hours of daylight. Many fish species feed most actively at first light of early morning, after fasting during the night. Automatic systems, switched on by photocells, can deliver feeds outside "normal" farm working times and take advantage of these active feeding periods.

Automatic feeding systems are also available which move food along pipes from central storage silos or mixing units. The timing and quantities of feed delivered to holding units are controlled by computerized systems capable of integrating environmental information (water temperature and light conditions) with fish biomass data (fish numbers and average weights) to calculate, control, and record feed quantities and feeding frequency. Remote feeding systems are capable of delivering moist or dry feed to as many as 32 holding units and can be installed on shore bases, barges, or boats. Food is most commonly transported through the pipes in a flow of pumped water, although blown air is used in some systems. The pipes are typically 50 mm in diameter capable of carrying a full range of pellet sizes for distances greater than 500 m. Piped feeding systems are used most commonly at sea cage sites in the farming of Atlantic salmon, rainbow trout, sea bass, and sea bream (Fig. 7.10).

Fig. 7.7 An automatic feed dispenser. Dry feeds are conveyed automatically from the hopper to the spreader. The range and pattern of distribution of the food are adjustable, and the frequency and quantities of feed delivered are preset to particular feeding regimens.

Some automated feeding systems available to fish farmers have been adapted from agricultural applications. Systems are available that distribute dry feeds in rigid pipes from storage hoppers to distribution points, from which the food is dropped or spread into fish holding systems. These systems move food along pipes by means of augers or by discs attached to moving cables (Fig. 7.11). These systems are adaptable for use on tank or raceway systems, and are used most frequently at indoor culture facilities, for example, in the culture of tilapia, eels, and striped bass.

Many automatic feeders used in cage culture are operated on the principal of frequent, small deliveries of food in order to achieve an even spread of food through the fish population. Most fish cages have a relatively small surface area in relation to the cage volume, and small frequent feedings may give more fish an opportunity to rise to the surface and feed compared with larger infrequent feedings. While this practice may be deemed necessary in heavily stocked cages, it may fail to accommodate the natural feeding behaviors of some species, and result in poor food conversion ratios. Continuous feeding disrupts the group structure of fish schooling in cages as fish continually compete for food. This pattern of behavior has been shown to reduce growth efficiency.

To accommodate the problems associated with feeding fish in cages hydroacoustic feed detectors have been developed which monitor food sinking through the cage. Since fish showing appetitive behavior rise to the surface of cages to compete for food, any food detected sinking below the top one-third of the cage may be contributing to wastage. In one series of experiments hydroaccoustic detection of food pellets which

Fig. 7.8 Compressed air feeders. Small dual feeders, here being used to feed Atlantic salmon smolts. The system is adjustable for the frequency of delivery and feed quantity.

had sunk to 2.5 m depth below the surface was taken as an indicator of reduced appetite (Juell et al., 1993). The release of food from automatic dispensers was stopped when the echo energy from food pellets sinking through a 360° acoustic beam exceeded a preset threshold (Fig. 7.12).

Results from feeding trials conducted with Atlantic salmon showed higher specific growth rates and lower food conversion ratios in groups of fish fed in cages with food detectors than in control groups fed in accordance with feeding tables based on growth rate estimates. All the fish were fed using automatic feeders which spread the food over an area of 16 m². The control groups were fed continuously at low intensity, while the experimental groups were fed four meals per day. Food intake in the experimental groups varied considerably from 0.24% to 3.93% of the estimated biomass. This re-

Fig. 7.9a   Large compressed air feeders. A deflector is attached to the end of the distribution pipe to ensure an effective spread of food.

Fig. 7.9b   A weighing balance is incorporated for adjustment of feed quantities. Photographs courtesy Trond Severinsen, Popfeeder (Canada) Inc.

Fig. 7.10   Feed distribution pipe. Moist feed is delivered to fish through a hydraulic piped system, from a central distribution point.

flected natural fluctuations in appetite which were accommodated by the demand feeding system.

Hatchery Feeders

Feeding systems used in hatcheries must be capable of delivering feed almost continuously since small fish are growing rapidly and have high metabolic demands. Hatchery feeds are generally the most expensive types of feed used in aquaculture but are used in relatively small quantities in proportion to the grower diets which are used throughout the major portion of the growing cycle. Considerable ingenuity has been applied to the design of hatchery feeders. Water driven, clockwork powered, and a variety of electrically powered types are available. Since hatcheries normally comprise large numbers of small holding tanks, centralized control systems are common.

Hatchery feeds are finely powdered, or crumbled, and may contain large proportions of oil, hence they may be sticky with poor flow characteristics. The nature of hatchery feeds has influenced the design of feeders. Many types comprise a revolving disc or plate on which food is placed, or fed from a hopper. As the plate slowly revolves, food is pushed from the plate by a radial arm, and falls into the tank below. By regulating the speed of rotation of the disc the quantity of food dispensed can be controlled

7.11 Automated feed delivery. Feed dispensers are refilled via rigid pipes from a storage hopper. The dry food is moved by a system of discs attached to a central cable (see right of photograph). Photo courtesy Cablevey Iowa, USA

reasonably accurately. Since the disc is rotating continuously a small constant measure of food is released into the tank (Fig. 7.13).

Another popular type of hatchery feeder incorporates a clockwork driven moving belt. A weighed quantity of food, to cover the day's ration, is spread along the belt. As the belt moves slowly forward, food falls from the edge of the belt through a wide opening into the tank below. Different sizes of belt feeder are available which are capable of releasing food continuously over an 8- or 12-hour period (Fig. 7.14). Belt feeders are relatively cheap, or easy to construct, and operate efficiently in damp hatchery conditions. Since feeding to slight excess is normal practice in fish hatcheries, water quality must be maintained by providing adequate water exchange to flush uneaten food from the rearing tanks.

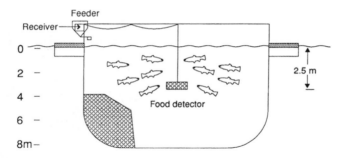

Demand feeding in salmon farming by hydroacoustic food detection

Fig. 7.12   Hydroaccoustic food detection. A demand feeding system for use in the cage culture of salmon. The automatic feeder is controlled via signals from a hydroaccoustic detector which monitors feed falling through the cage. From Juell et al., 1993.

Mobile Feeders

Mobile feeders combine some of the advantages of hand feeding with the labor saving benefits of automated feeders. At their simplest mobile feeders comprise a hopper and compressor-driven blower which can be mounted on a truck or boat. They can also be connected to the power takeoff of farm tractors (Fig. 7.16). Using a flexible broad-

Fig. 7.13   A revolving plate feeder. Feeders of this type are widely used in the hatchery production of fry and fingerlings.

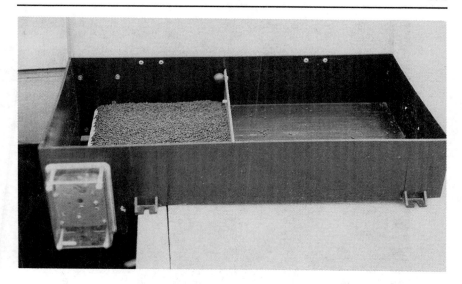

Fig. 7.14 A belt feeder. Clockwork-driven feeders are a cheap alternative to externally powered systems.

cast hose an experienced worker can control the quantity of feed released into each holding unit in accordance with the feeding response of the fish, or can feed a predetermined amount of food. Mobile systems for use on boats are available that use either blown air or pumped water to propel the food. Mobile feeders are available in a wide range of sizes, most of which can be operated by a single person (Fig. 7.15).

Subsurface Feeders

A novel approach to avoiding feed waste in cage-rearing systems is that of releasing extruded floating food from devices at the bottoms of cages (Jaffa, 1994). Food, of selected density, released from the floor of the cage, slowly floats upward through the cage toward the surface. Uneaten food, arriving at the surface, indicates the need to reduce or stop feeding. Subsurface feeding increases the time available for food consumption, since food will float at the surface for a period of time dependent on wave and current action. It also reduces the need for fish to make regular migrations to the cage surface seeking food. These systems are currently undergoing farm trials (Jaffa, personal communication).

FEEDING METHODS FOR SHRIMP

Feeding shrimp in ponds poses a number of special problems, and necessitates particular understanding of the effects of water quality and other environmental factors on

Fig. 7.15 A water-powered mobile feeder. These feeders broadcast food in a water jet, and can be operated from cage walkways or from boats. Photo courtesy Joanne Constantine.

feeding behavior. A number of critical factors should be taken into account when planning feeding strategies.

Feeding Behavior

Shrimp are less mobile than fish and do not aggregate in response to the presence of feed in a manner comparable with intensively farmed fish. This necessitates careful and even distribution of feed. Patterns of distribution of shrimp in ponds also differ de-

Fig. 7.16 An air-powered mobile feeding unit. A tractor-towed pneumatic feeder in use at a striped bass farm in Louisiana, USA.

pending on their stage of development. Early juvenile stages of most species prefer to feed in the shallower water along the walls of the dike. As development proceeds through late juvenile into the adult stages the shrimp feed freely on the benthos of the whole pond. Hence distribution of the feed for the early juvenile stages may be most effectively carried out from the dike, while for later stages feed should be distributed over the whole pond combining feeding from boats and from the dike. During feeding attention should be paid to the direction of the wind across the pond (Fig. 7.17). Shrimp congregate in areas where there is turbulence due to wind action and more feed should be distributed to these areas. Some shrimp species forage more actively at night than during the day, and there is also some evidence that diurnal feeding patterns vary with the age of the shrimp. Tidal and lunar cycles may also influence patterns of feeding behavior in shrimp ponds.

Normal feeding patterns are interrupted during molting. Immediately before, and during the actual molt the shrimp stop feeding. This typically lasts for 1–3 days after which feeding levels increase above normal. Feeding levels will need adjustment if a significant percentage of animals molt simultaneously. Such mass moltings may be triggered by large-scale water exchanges. The degree of molting activity can be estimated by observing shrimp on feeding trays.

Distribution of shrimp in ponds is also influenced by various elements of water quality. During midday, when water temperatures are highest, shrimp will move to forage in the deeper, cooler regions of the pond. Shrimp will also move away from any ar-

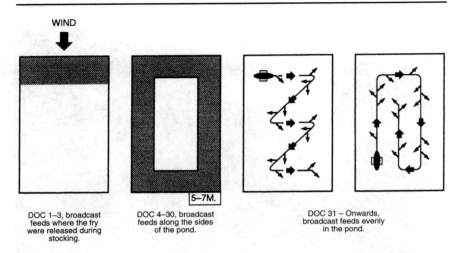

WIND

DOC 1–3, broadcast feeds where the fry were released during stocking.

5–7M.

DOC 4–30, broadcast feeds along the sides of the pond.

DOC 31 – Onwards, broadcast feeds evenly in the pond.

Fig. 7.17 Feed dispersion at different stages of semi-intensive shrimp culture. Reproduced with permission of San Miguel Foods Inc., Manila, Philippines. DOC = Day of culture.

eas where dissolved oxygen levels are low, or where there is an accumulation of ammonia or hydrogen sulphide.

There are often accumulations of silt, uneaten food, and feces during the latter stages of each shrimp production cycle. This is particularly significant in intensive systems and where water exchange rates are poor. The current-generating actions of paddlewheel aerators, in fixed positions, can exacerbate this problem causing localized accumulations of organic waste which rapidly become anoxic. These areas will again be avoided by shrimp.

These factors, outlined above, influence the normal distribution of shrimp in ponds and must be taken into account during feeding. Shrimp are generally fed by hand, either from boats or from the dike. Mobile feeders may also be used from boats or trucks where the strength and width of the dike permits.

Feeding Trays

The use of feeding trays is widespread in shrimp aquaculture to enable direct observation of the feeding activity of the shrimp. A percentage (usually 1–3%) of the daily ration allotted to the pond is spread between a number of feeding trays. Feeding trays consist of a wooden or metal frame across which is stretched a fine mesh net. The frame dimensions vary between 70 cm × 70 cm to 1 m × 1 m (Fig. 7.18). The numbers of feeding trays used varies considerably between farms. Table 7-1 gives recommendations for numbers of feeding trays. Feeding trays rest on the pond bottom and may be serviced from walkways, the pond dikes, or from boats. Food is placed on trays at the

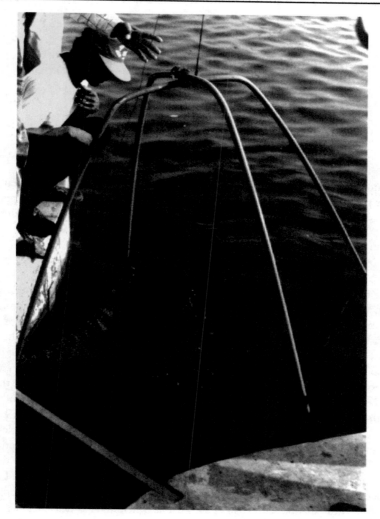

Fig. 7.18 A shrimp feeding tray.

same time that food is broadcast across the pond. After several hours the trays are care-fully lifted and an estimate made of the quantity of food remaining. The next feeding ration is then adjusted in light of the average amount of food remaining on the trays of each pond. Interpretation is subjective and shrimp farmers differ in their routine ad-justments to feeding levels (Chapter 8).

Feeding trays should be installed immediately after stocking to permit observation of the condition and growth of the shrimp, and to detect the presence of any pests or

Table 7-1   Recommended Number of
Feeding Trays for Different Pond Sizes

Pond Size	No. of Feeding Trays
0.5 ha and below	4
0.6 – 0.7 ha	5
0.8 – 1.0 ha	6
2.0 ha	10 – 12

predators. In positioning trays, areas should be avoided which are close to aerators or where the pond bottom is uneven or sloping, since food may be washed or tipped off the trays. Areas should also be avoided where there are localized buildups of sediments.

FEEDING METHODS AND GROWTH VARIATION

In the intensive culture of most fish species, variations in growth within single stocks occur. While this, in part, may be due to genetic differences a significant component of the variation may result from competitive interactions between fish. The level of interaction will vary between species, at different developmental stages, and for the same species held at different stocking densities. A consequence of competitive interactions is that some fish will outcompete others for food and will consequently grow faster, further exacerbating the problem. While regular grading of fish stocks will restore the size balance, there is also some evidence that choice of feeding method may influence growth patterns in farmed fish. Experiments with salmon (Ryer and Olla, 1991; Thorpe et al., 1990) have shown that where fish are able to establish defensive territories near food sources, aggressive interactions occur, which result in uneven consumption of food among the fish (Fig. 7.19). This situation is most likely to occur where fish are held at low densities. An even spread of food will result in more uniform access. In contrast localized mechanical delivery has been shown to result in greater competition. Experiments with Atlantic salmon have shown that where food is delivered at a point source in a seacage, most pellets are consumed by a small proportion (25%) of the population.

Waste Detection

Careful hand feeding, or the use of hand-operated mobile feeders, in culture systems with good water visibility should minimize the risks of overfeeding fish stocks to the point where food remains uneaten and is wasted. Avoiding wastage where automated feed delivery systems are used is more problematic, particularly in seacages, where the observable surface area is small relative to the volume of the cage, and where visibility

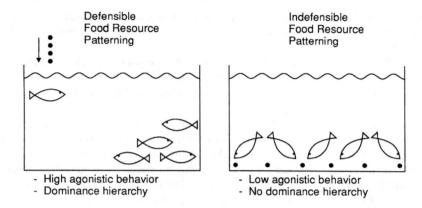

Fig. 7.19 A comparison of the effects of defensible vs. indefensible food resource patterns on the agonistic behavior of juvenile chum salmon. From Olla et al., 1990.

may be poor. In these situations control systems are being developed which are capable of detecting feed pellets if they fall through the bottom of the cage uneaten. This information can then be used to restrict any further feeding until the appetite of the caged fish returns. While these systems are not in common use, a number are available, and others are being developed. Uneaten food can be detected in a number of ways. It can be observed directly using submerged video cameras linked to surface monitors; it can be detected using hydroacoustic systems; or collected in a trap below the cage, the contents of which are continuously pumped, or airlifted to the surface during feeding. These techniques can only be applied at cage farm sites with low or predictable water currents (Juell, 1991). At more exposed sites feed may be swept though the vertical sides of the cage and remain undetected. Hydroacoustic monitors are being developed with feedback systems which can automatically regulate the operation of feed dispensers. If such systems can be developed at an acceptable cost they would have a real potential to deliver feed in close accord with the varying appetite and food intake of caged fish populations.

REFERENCES

ADRON, J.W., P.T. GRANT and C.B. COWEY. 1973. A system for the quantitative study of the learning capacity of rainbow trout and its application to the study of food preferences and behaviour. Journal of Fish Biology 5:625–636.

ALANÄRÄ, A. 1992a. Demand feeding as a self-regulating feeding system for rainbow trout (*Oncorhynchus mykiss*) in net pens. Aquaculture 108:347–356.

ALANÄRÄ, A. 1992b. The effect of time restricted demand feeding activity, growth and feed conversion in rainbow trout (*Oncorhynchus mykiss*). Aquaculture 108:357–368.

BOUJARD, T. and J.F. LEATHERLAND. 1992. Demand-feeding behaviour and diel pattern of feed activity in *Oncorhynchus mykiss* held under different photoperiod regimes. Journal of Fish Biology 40:535–544.

CRAMPTON, V., H. KREIBERG, and J. POWELL. 1990. Feed cost control in salmonids. Aquaculture International Congress Proceedings, 99–108. Aquaculture International, Vancouver.

CRUZ, P.S. 1991. Shrimp Feeding Management—Principles and Practices. Kubukiran Enterprises Inc., Davao City, Philippines.

ERIKSON, L.O. and A. ALANÄRÄ. 1990. Timing of feeding behaviour in salmonids. In The Importance of Feeding Behaviour for the Efficient Culture of Salmonid Fishes, eds. J.E. Thorpe and F.A. Huntingford, pp. 41–48. World Aquaculture Society, Baton Rouge.

GROVE, D.J., L.G. LOZOIDES, and J. NOTT. 1978. Satiation amount, frequency of feeding and gastric emptying rate in *Salmo gairdneri*. Journal of Fish Biology 5:507–517.

JAFFA, M. 1994. Sub-surface feeding—an answer to waste. Fish Farmer Jan./Feb. 51.

JUELL, J.E. 1991. Hydroacoustic detection of food waste—a method to estimate maximum food intake of fish populations in sea cages. Aquaculture Engineering 10:207–217.

JUELL, J.E., D.M. FUREVIK, and Å. BJORDAL. 1993. Demand feeding in salmon farming by hydroacoustic food detection. Aquaculture Engineering 12:155–167.

LANDLESS, P.J. 1976. Demand-feeding behaviour of rainbow trout. Aquaculture 7:11–25.

OLLA, B.L., M.W. DAVIS and C.H. RYER, 1990. Foraging and predator avoidance in hatchery-reared Pacific salmon: achievement and behavioural potential. In: The Importance of Feeding Behaviour for the Efficient Culture of Salmonid Fishers, eds. J.E. Thorpe and F.A. Huntingford, pp. 13–20. World Aquaculture Society, Baton Rouge.

RYER, C.H. and B.L. OLLA. 1991. Agonistic behaviour in a schooling fish. Environmental Biology of Fishes 31:355–363.

SINGH, T. 1993. Feeding systems in aquaculture. Infofish International 2/93:44–48.

THORPE, J.E., C. TALBOT, C.E. RAWLINGS, M.S. MILES and D.S. KEAY. 1990. Food consumption in 24 hours by Atlantic salmon (*Salmo salar L.*) in a sea cage. Aquaculture 90:41–47.

8

`\ \ \ \ \ \`

Feed Rations and Schedules

INTRODUCTION

The selection of appropriate ration sizes and feeding schedules are critical elements in effective feed management. The size of the daily food ration, and the frequency and timing of meals, are key factors influencing growth and food conversion. While some farmers feed to a set pattern with little variation, the estimation of daily rations should take into account the numerous factors, biotic and environmental, which influence appetite. Ration sizes should accommodate the short-term fluctuations which occur in appetite, and should be adjusted to meet the changing feed demands as fish grow larger, and water temperatures and other environmental factors change. They should also be adjusted, as necessary, to support the production and harvesting objectives of the farm.

RATION SIZE

In selecting the ration, or daily food allotment, a number of basic choices are possible. Food can be provided in excess, to apparent satiation, or in restricted amounts.

Excess

Feeding to excess, or *ad libitum*, implies a constant availability of food. In conventional livestock production, feeding to excess is a common option since any unconsumed food can be removed, measured, and offered again. In contrast, feeding to excess in aquatic systems is extremely wasteful, since once wetted the food begins to denature, is generally inaccessible, and cannot be reused. Feeding to excess leads to poor food conversion, and the disintegration of uneaten food causes degradation of water quality. Only in the early stages of hatchery rearing is feeding to *slight* excess an accepted practice.

Satiation

Feeding to satiation involves feeding fish the maximum amount of food that they will consume. In practice, feeding to satiation, while avoiding wastage, may be difficult to achieve using many of the available feeding systems described in Chapter 7. Farmers seeking to feed fish to satiation typically feed the maximum amount of food which the fish will consume several times a day, the actual number of meals depending primarily on the size of the fish and on water temperatures. For carnivorous species this practice results in close to maximum ingestion levels and growth rates.

Restricted

Restricted rations are predetermined ration sizes set below the estimated maximum ration size. While fish farmers are generally seeking to maximize the growth rates of fish or shrimp, there may also be periods in farm production cycles where the objectives are to reduce or control growth rates as part of planned production strategies. There is some evidence from feeding and growth experiments that feeding restricted rations permits greater control of food conversion ratios. Where maximum growth is the objective, farmers must seek to establish feeding regimens in which fish or shrimp consume food at levels of, or approaching satiation, while at the same time avoiding wastage.

RATIONS AND GROWTH

An understanding of the relationships between growth, rations, and food conversion is fundamental in optimizing the use of feeds in intensive fish culture. Most information has been derived from experimental work on salmonids (Brett and Groves, 1979), although ration and growth experiments have also been conducted with other species, including common carp (Huisman, 1976; Bryant and Matty, 1981), and the African catfish, *Clarius gariepinus* (Hogendoorn, 1981a,b); and tilapia hybrids (Clarke et al., 1990).

Generalized examples of a growth-ration curve and a food conversion-ration curve are illustrated in Fig. 8.1a,b. From the growth-ration curve it can be seen that fish lose

(a)

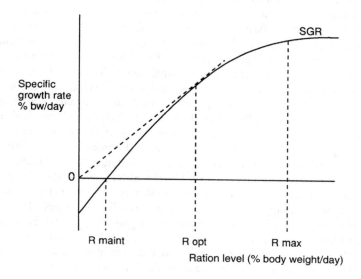

(b)

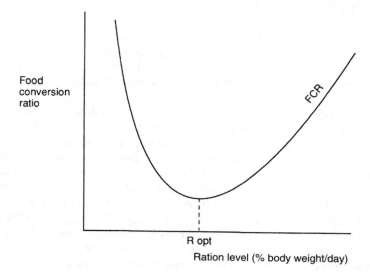

Fig. 8.1 (a) Effect of feeding rate on growth rate. (b) Effect of feeding rate on food conversion.

weight when their food intake falls below that required for maintenance (R_{maint}). As ration size and food intake increase, the growth rate increases up to the point of maximum voluntary food intake (R_{max}) and maximum growth rate. The ration size corresponding to the point of inflexion of the curve is often expressed as the optimum ration (R_{opt}) (Brett et al., 1969), since this represents the point of maximum conversion efficiency, or the point at which growth per unit of food intake is greatest. From Figures 8.1(a) and (b) it can be seen that maximum growth and the lowest food conversion rates do not coincide at the same ration level. Maximum conversion efficiency occurs at ration levels below those at which maximum growth occurs. From these results it is evident that there is a range of possible feeding levels, the choice of which depends on whether maximum growth, optimal food conversion, or a balance between the two is sought.

With improvements in the formulation and performance of aquaculture feeds, some reexamination of the growth-ration data on which many current feeding recommendations are based may be necessary. The use of high energy diets in trout and salmon culture stimulates appetite and results in increased growth and feed conversion efficiency. It has been suggested that under optimal conditions a straight-line relationship exists between ration level and growth up to the point of R_{max}, and that maximum growth and food conversion efficiency both occur when fish are fed maximum rations (Talbot, 1994a). In practice many farmers feed to maximum ration for maximum growth. The economic gains of increased production should however be measured against any increased costs incurred in feeding to maximum ration.

GROWTH–RATION CURVES

A growth–ration curve, based on data collected from feeding a range of ration sizes, can provide a useful feed management tool, accepting that it will be specific for any particular species and size of fish, type of food, and husbandry system. A growth–ration curve can be used to examine the economic implications of feeding at different ration sizes.

A growth–ration curve can be derived by taking time and dedicating some production space to a relatively simple feeding experiment. This entails stocking 5 ponds, tanks, or cages with equal numbers of fish of the same size. The most useful information will be gained by using fish in the final stage of their production cycle, since during this period, food intake levels are generally highest and fish are growing rapidly.

Fish in each group should be accurately weighed and counted to give an estimate of total biomass. Each group of fish is assigned a ration size ranging from 50–100% of maximum ration (e.g., 50% R_{max}, 60% R_{max}, 67% R_{max}, 75% R_{max}, and 100% R_{max}). If fish are fed meals of reduced size on a daily basis, then larger or more aggressive fish may feed at the expense of smaller fish. Larger fish may feed at R_{max}, while others within the same group feed are feeding on restricted rations. This problem can be reduced by cutting back on the number of meals rather than reducing meal size. A suggested protocol is as follows (Furnell, 1990):

- Group 1 Two meals every day   = 100% Rs
- Group 2 Six days feed/two days starve   = 75% Rs
- Group 3 Four days feed/two days starve = 67% Rs
- Group 4 Three days feed/two days starve = 60% Rs
- Group 5 Two days feed/two days starve   = 50% Rs

At each meal fish should be fed to satiation and a record kept of the food consumed by each group.

The data collected from an experiment of this type can be used to plot a growth–ration curve. From the point of intersection of the tangent to the curve the optimal ration (R_{opt}) can be calculated. Information on growth rates and conversion values can be used to assess feeding strategies, in particular, to examine the financial implications of feed ration size (see Chapter 10). The strategy outlined above is drawn from experience with cage culture of chinook salmon (Furnell, 1990) and may require modification before being applied to other species. Fish requiring more frequent feedings than salmonids may not respond to periods of starvation. For these species it may be necessary to feed reduced amounts on a daily basis, which may result in increased size variation necessitating size grading during the feeding trials.

ESTIMATING RATION SIZE

Various methods are used to estimate ration size. These include calculations based on the use of feeding charts, feed equations, and growth predictions. The methods used vary for different species and culture systems and in practice many farmers use a combination of methods to determine daily feed requirements.

Feeding Tables

Feeding charts were first developed as guides to the daily feeding requirements of hatchery-reared salmonids (Deuel et al., 1952). The daily food rations given in these tables were calculated from estimates of biomass, food conversion, and expected growth. Tabulated feeding levels for various species have been progressively modified in line with the improvements which have been made in feeds and feeding methods, and tables are now available for most species of intensively farmed fish and shrimp. They are commonly supplied by feed manufacturing companies and give values for daily food rations, expressed as a percentage value of the total weight or biomass of fish to be fed daily (Table 8-1). Ration sizes are generally cited described by the term %body weight/day; practically expressed as kg food/100 kg biomass/day. Most tables relate daily food rations to two factors: fish size and water temperature. As discussed in Chapter 2, the percentage of food consumed in relation to body weight decreases as the fish grow larger and their metabolic demands decrease. Food intake is also directly related

Table 8-1 Feeding Guide for Salmonids in Freshwater. Figures Give the Feeding Rate in Percent of Body Weight Per Day, (the Weight of Feed in kg Per Day for Each 100 kg of Fish

Fish Weight	-0.5g	0.5–1.2g	1.2–5g	5–12g	12–25g	25–40g	25–40g	40–80g	80g+
Fish Size	-1:	1–1.5"	1.5–2.5"	2.5–3.5"	3.5–4.5"	4.50–6"	4.5–6"	6–8"	8"+
Feed Size	0.5mm[1]	0.7mm	1.0mm	1.5mm	2.0mm	3.0mm	2.5mm pt[2]	3mm pt	4mm pt
Water Temp									
2	2.7	2.2	1.7	1.3	1.0	0.8	0.7	0.6	0.4
4	3.2	2.6	2.2	1.7	1.3	1.0	1.0	0.8	0.5
6	4.1	3.0	2.5	1.9	1.4	11.2	1.2	0.9	0.6
8	5.3	3.7	2.8	2.0	1.7	1.4	1.4	1.0	0.7
10	6.0	4.3	3.4	2.2	2.0	1.7	1.7	1.2	0.9
12	6.3	4.9	3.9	2.6	2.3	1.9	1.9	1.4	1.0
14	6.5	5.5	2.9	2.6	2.1	2.1	2.1	1.5	1.1
16	7.8	6.5	5.3	3.1	3.1	2.5	2.3	1.8	1.3
18			5.3	3.5	3.5	2.8	2.5	1.9	1.5
20			5.9	4.0	4.0	3.2	3.0	2.1	1.7

[1]Granules

[2]Pellets

Reproduced with permission of Corey Aquaculture, Fredericton, New Brunswick, Canada

to temperature, increasing with temperature so long as temperature remains within the fishes' optimal range.

Feeding charts provide only a general guide to feed intake and do not take into account either the short- or long-term fluctuations which occur in appetite in response to numerous physiological and environmental factors. The daily feeding rations given in charts generally represent maximum food intake, consistent with efficient digestion and uptake of food under optimal conditions. Under practical farming conditions they should be used to provide only an approximate guide to daily food intake. They are however useful for estimating feed requirements. For many farmers they provide baseline values which are subsequently modified as feeding and growth data are collected from the various batches of fish growing on the farm.

There is a current need for improved feeding charts that reflect the improvements in growth performance of farmed fish stocks. Some updating of feeding tables has been made possible from controlled growth studies carried out at those research centers with facilities for farm-scale trials. Results derived from feeding experiments conducted in Norway with rainbow trout and Atlantic salmon reflect higher growth rates than were previously accounted for in published feed tables (Austreng et al., 1987). Tables derived from these studies incorporate data gathered under controlled conditions over a 5-year period. The tables list growth rates, expressed as %weight/day, and are directly applicable as feeding guides, since under the practical experimental conditions, food conversion ratios of approximately 1:1 (Food fed:weight gain) were obtained (Table 8-2).

While most published feeding tables are based on predictions of growth and feed conversion, more recently methods have been proposed, and feeding tables published, for estimating the daily rations of rainbow trout, based on the known requirements of this species for digestible energy and nutrients. Information for farmers of this type is presently restricted to those few species for which digestible energy requirements have been established, across appropriate ranges of fish size and water temperature. Estimating feed requirements from known energy and nutrient requirements forms the basis of feeding systems for most types of farmed livestock. As more detailed and reliable infor-

Table 8-2a Growth Rate (% wt/day) of Young Trout in Fresh Water

Water Temperature (°C)	Weight (g)							
	0.12–0.5	0.5–1.0	1.0–2.5	2.5–5	5–10	10–20	20–40	40–100
4	5.0	3.8	3.3	2.3	1.5	1.2	0.9	0.7
6	6.0	4.5	3.7	3.0	2.3	1.9	1.6	1.3
8	7.0	5.5	4.2	3.5	3.0	2.5	2.2	1.9
10	7.5	7.0	5.5	4.1	3.5	3.0	2.6	2.3
12	8.0	8.5	6.0	4.5	4.0	3.5	3.0	2.7
14	8.5	9.5	7.0	5.5	4.5	4.0	3.5	3.1
16	9.0	10.5	8.0	6.5	5.5	4.5	4.0	3.5

Table 8-2b   Growth Rate (% wt/day) of Atlantic Salmon and
Rainbow Trout in Sea Cages

Water Temperature (°C)	Weight (g)			
	30–150	150–600	600–2000	>2000
2		0.2	0.2	0.1
4		0.5	0.3	0.2
6		0.7	0.5	0.3
8	1.3	1.0	0.6	0.4
10	1.6	1.2	0.8	0.5
12	1.9	1.4	1.0	0.6
15	2.2	1.7	1.1	0.7

From Austreng, et al., 1987

mation becomes available for cultured fish species, this approach to estimating ration sizes will undoubtedly become more widely used in aquaculture.

Feed Calculations

Feed calculations are made on the basis of the interrelationships between ration size, growth, and food conversion:

$$\text{Daily Ration}$$

$$\text{Food Conversion} \underset{\displaystyle \text{(FCR)}}{\overbrace{\hspace{4cm}}} \text{Daily Growth Rate} \atop \text{(SGR)}$$

With knowledge or estimates of growth rates and predictions of food conversion values daily rations can be calculated where:

$$\text{Daily Rations} = \text{SGR} \times \text{FCR}$$

Feed calculations are specific to particular culture operations and involve the calculation of hatchery constants. A method devised by Haskell (1955) has been widely applied in trout and salmon culture.

$$\text{Daily ration (\%body weight/day)} = dL \times FCR \times 3 \times Ld$$

where:

dL = daily length increase
FCR = food conversion ratio
Ld = length of fish on day of feeding
To use this equation it is first necessary to establish the length of an appropriate

growth period. Fourteen days is commonly used, and the anticipated daily length increase over this period is determined from previous records. The daily length increase is then determined by dividing this length increase by 14. An anticipated food conversion value is obtained from previous records, and the respective values are used in the equation to calculate daily ration. From Day 2 onward the Ld value is increased each day by the dL. This ensures that farmers are feeding not just the original biomass, but each additional daily increment of growth. Where the water temperature, diet, and species remain constant the factors in the numerator of the equation also remain constant, and can be combined into a hatchery constant (HC)

where:

$HC = 3 \times conversion \times dl \times 100$
Daily ration $= HC/L$

Experimental Estimations of Food Intake

With increasing interest in optimizing feeding rates various approaches are being taken to establish food intake levels using direct experimental methods. Methods have included starving fish for short periods, feeding to excess, and then killing and dissecting the fish in order to determine the weight of food in their stomachs. Other techniques, which can be used on live fish, include feeding fish on diets containing either radioisotopes (Storebakken et al., 1981) or low concentrations of X-ray dense particles (Talbot and Higgins, 1983; Jobling et al., 1993). Both of these methods enable estimations to be made of the stomach contents of fish feeding to excess, and hence give values for maximum food intake levels. Information of this type may be used to devise or modify feeding tables.

FOOD PARTICLE SIZE

The maximum particle size that fish and shrimp can consume increases as the animals grow. Experiments have shown that food particle size can affect growth and feed conversion efficiency, and that optimal food particle sizes can be identified for fish and shrimp of different sizes. Studies on a number of fish species have examined the relationships between food particle size, mouth size, fish size (fork length), and weight gain. In marine fish larvae, optimal food size is about 25% of mouth width at first feeding, and this increases to about 50% within a few days. The optimal width of prey for most marine fish larvae which have been cultured is within the range, 35–100 microns (Tucker, 1992). For the European eel, Anguilla anguilla, optimum particle size is 40–60% of mouth size (Knights, 1983), while in experiments with juvenile Atlantic salmon, feeding particles 25% of mouth width gave maximum increase in weight and length (Wankowski and Thorpe, 1979). In a detailed study on the feeding of Arctic

Table 8-3 Recommended Pellet Sizes for Juvenile and Adult Tilapia

Fish Size/Age	Particle Size (diameter)
Fry: first 24 h	Liquifry
Fry: 2nd day–10th day	500/μm
Fry: 10th day–30th day	500–1000/μm
Fry: 30th day–juveniles of 0.5g–1.0g	500–1500/μm
1g–30g	1–2mm
20g–120g	2mm
100g–250g	3mm
250g+	4mm

From Jauncey and Ross, 1982

charr, Tabachek (1988) recorded the highest growth rates for 3–7 g fish fed particles 23–25% of mouth size, for 9–12 g charr fed particles 23–25% of mouth size, and for 16–21 g charr fed particles 31–33% of mouth size. These results illustrate the specific requirements of fish for particle size. Feed particles must be of a size that can readily be ingested. While large fish can ingest small particles, it requires more energy for a fish to capture an equivalent weight of smaller particles. This results in a measurable reduction in food conversion efficiency. An example of recommended pellet sizes for juvenile and adult tilapia is given in Table 8-3 (recommended pellet sizes for rainbow trout are shown in Table 8-1).

In contrast to the situation in fish, shrimp feed pellet size is not related to the mouth size of the shrimp. As described in Chapter 2, shrimp physically break up pellets prior to ingestion. The pellets must, however, be of an appropriate size so that the shrimp can readily carry the pellets while swimming and manipulate them with their feeding appendages. Shrimp feeds are generally manufactured in three pellet sizes (recommended shrimp pellet sizes are given in Table 8-4).

A number of factors will influence the selection of appropriate feed sizes. There may be a lack of availability of a full size range of diets for a particular species. This may be limited by regional supplies of feed or limited size ranges produced by local manufac-

Table 8-4 Recommended Shrimp Pellet Sizes

Feed Type	Size of Shrimp (gm)	Feed Size
Starter	0–3.0	1–2mm crumble
Grower	3.0–15.0	2.0–2.5mm x 4.0–5.0mm
Finisher	15.0–40.0	2.0–2.5mm x 6.0–8.0mm

From Akiyama, 1993

turers. This is not uncommon in obtaining feeds for fry and fingerlings. Since these fish are growing rapidly they require a wide range of sizes over a relatively short space of time. This necessitates careful planning and advance purchase of sufficient quantities of the appropriate feed types and sizes. Care must also be taken to select the appropriate feed sizes in situations where feeds manufactured for a particular species are used to feed another species, for example, in the use of trout feeds in tilapia or striped bass culture.

Appropriate feed sizes can only be selected where stocks of fish are adequately size graded. If a wide range of fish sizes are present they would require feeding with an equivalent range of particle sizes. In practice this would be difficult given that the larger fish in the group will readily consume the smaller particles. To optimize growth necessitates regular size grading of stocks (Fig. 8.2). To accommodate the normal size range variation found in all fish stocks, particle sizes should be blended in feeds during the switch-over from one particle size to another.

Pellet Shape

In addition to pellet size some recent attention has been given to pellet shape. Fish pellets used in grow-out operations are generally cylindrical in shape, with relatively precise diameters and variable length. Some recent studies have shown that shorter pellets are more readily ingested by Atlantic salmon than longer pellets (Smith et al., 1995). A new shape of commercial feed has also been introduced recently. These have

Fig. 8.2 Transferring and size grading Atlantic salmon.

a somewhat flattened cylindrical shape, which gives the pellets a characteristic motion as they fall through the water column. The manufacturers claim that this is more attractive to fish than the trajectory of a conventional pellet. It can be anticipated that more attention will be focused in the future on any possible gains to be made by using different shapes, colors, and textures of pellets on ingestion rates (Strademeyer, 1992).

FEEDING FREQUENCY

The daily food ration may be fed as a single meal or more typically is divided into a number of separate feeds spread throughout the day. The optimal feeding frequency for fish or shrimp may vary depending on species, age, size, environmental factors, and food quality. Optimal frequencies have been reported in fish, which range from continuous feeding in African catfish fry, *Clarias gariepinus* (Hogendoorn, 1981a), to one meal every other day in young estuary grouper, *Epinephelus salmoides* (Chua and Teng, 1978), and seabass, *Dicentrarchus labrax* (Carillo et al., 1986). From shrimp feeding studies, daily rations divided into four meals per day have been reported optimal for the banana shrimp, *Penaeus merguiensis* (Sedgewick, 1979), and the white Pacific shrimp, *Penaeus vannamei* (Robertson et al., 1993).

Fish

The limited information available suggests that optimum feeding frequencies can be determined for each species and for different sizes of fish within a single species. Siraj et al. (1988), fed hybrid tilapia fry a 10% body weight/day food ration divided into various meal schedules ranging from one meal every other day to three meals per day, while a control group was fed to excess. The optimal feeding frequency was twice per day. At this frequency the lowest food conversion ratios were observed and the growth rates were highest. An optimal feeding frequency for channel catfish of twice per day was also reported, following feeding trials at frequencies ranging from one to 24 meals per day (Andrews and Page, 1975).

Shrimp

Shrimp are generally fed 3–5 times each day. Controlled experiments have shown that increasing the feeding frequency from one to four meals each day, improved both feed conversion and growth rates of *Penaeus merguiensis* (Sedgewick, 1979), and growth rates of *Pennaeus vannamei* (Robertson et al., 1993). The stability of shrimp feeds in water may also be a factor in determining optimal feeding frequencies. If the use of less stable feeds cannot be avoided, they should be fed in smaller, more frequent meals.

FEEDING PERIOD

There have been relatively few studies of the timing of daily feeding activity of fish or shrimp. Considerable diversity is evident, however, and poses the question as to whether it may be advantageous, in terms of growth and food conversion, to feed fish and shrimp in accordance with their natural feeding patterns, as opposed to imposing artificial farm routines.

Fish

Food intake and growth in fish are generally linked to two main factors: water temperature and fish size. However, studies of feeding patterns in fish, given unlimited access to food, have also shown that feeding activity may follow both daily and seasonal fluctuations linked to photoperiod.

Most studies of rhythmicity of feeding have been conducted with salmonids. Seasonally changing, daily rhythms of feeding have been recorded in rainbow trout, with wintertime peaks of feeding activity occurring after sunset, and summertime peaks either after sunrise or in the middle of daylight hours. In a further study with rainbow trout, it was observed that 42% and 48% of the daily feeding demand occurred during the 4h period following dawn, under controlled photoperiods of 16L:8D and 12L:12D respectively (Boujard and Leatherland, 1992).

Adjusting feeding schedules to the assumed daily feeding rhythms for this species (one and two diurnal peaks) failed, however, to demonstrate any gains in growth rate or food conversion, when compared with continuously fed control groups of fish (Mäkinen, 1993). These latter observations support the view that salmonids, like other farm animals, are able to adapt physiologically and behaviorally to artificial feeding patterns entrained by food availability (Talbot, 1994b).

Shrimp

There is little information available about the time when penaeids feed. Most farmed shrimp are grooved species which are active at night and burrow in the substrate for at least part of the day. Suggested feeding regimens reflect these natural feeding habits, with most food given in the evening period. A proposed feeding schedule is shown in Table 8-5. *Penaeus japonicus*, are generally fed their full rations in several meals spread through the night. It has been reported that digestive enzyme activity in juvenile kuruma shrimp peaked 2–3 hours after dark, and that better growth rates resulted from feeding during the hours of darkness (Cuzon et al., 1982). Detailed observations of diurnal feeding activity have also revealed changing feeding patterns in relation to age for this species. Reymond and Lagardère (1990) found that 0.5 g shrimp fed during both day and night, 3 g animals fed more at night than during the daytime, while 7 g animals fed almost exclusively at night.

Table 8-5   Shrimp Feeding Schedule

Feed Type	Shrimp Weight	Feeding Time				
		0600	1000	1400	1800	2200
Starter	up to 3gm	30%	—	35%	—	35%
Grower	3–15gm	20%	15%	15%	30%	20%
Finisher	more than 15gm	20%	15%	15%	30%	20%

From Akiyama, 1993

The few reports available on the patterns of feeding behavior of farmed shrimp indicate considerable diversity, and the need for further research is indicated. Complex feeding regimens based on current information have been devised which take into account changes in diurnal feeding activity as shrimp grow (Abesamis, 1989). In large, semi-intensive and intensive systems the logistics necessary to implement nocturnal feeding may be impractical or uneconomical (Clifford, 1992). A recent experimental study with *Penaeus vannamei* showed that the timing of feeding (day vs. night vs. day and night) for this species had no effect on growth (Robertson et al., 1993).

Feeding trays provide the best mechanism for determining ration size in shrimp ponds. Shrimp appetites vary due to changes in environmental conditions such as water quality, water temperature, sunny or overcast days, and physiological factors such as disease and molting. These factors, along with the variable availablility of natural prey organisms, complicate the use of feed guides and tables. A feed guide for the cultured shrimp in semi-intensive culture is given in Table 8-6. Of the recommended feed amounts, a percentage of food should be allocated to feeding trays at each feed (Table 8-7). Based on the consumption levels of food in the trays, within the times specified for different sized animals, daily food allocations can be adjusted to match feed consumption levels (Table 8-8). While the use of feed trays is subjective, their careful and consistent use will help avoid the problems associated with over- and underfeeding shrimp stocks. Recommended feed frequencies and feed particle sizes usually are incorporated into commercial feed charts as guides for customers.

COMPENSATORY FEEDING

Compensatory growth is a period of greater than normal growth, following a period of growth restriction caused by undernutrition. The occurrence of compensatory growth is well known from growth studies on animals which, following temporary food deprivation, grow rapidly to make up any loss. In wild fish, compensatory growth occurs during the post-spawning period as depleted energy reserves are restored, and is also evident in many temperate species which feed voraciously during the early spring after a winter period of restricted food availability.

Table 8-6   Daily Feed Allotment Per 100,000 Shrimp With 80% Survival Rate

Weight (gm)	Biomass (kg)	Feed Rate (%)	Daily Quantity (kg)
up to 10 days	—	—	4
10–20 days	—	—	8
20–30 days	—	—	12
3	240	5.7	14
4	320	5.4	17
5	400	5.1	20
6	480	4.8	23
7	560	4.6	26
8	640	4.4	28
9	720	4.21	30
10	800	4.0	32
11	880	3.9	34
12	960	3.7	36
13	1040	3.6	37
14	1120	3.5	39
15	1200	3.3	40
16	1280	3.2	41
17	1360	3.1	42
18	1440	2.9	42
19	1520	2.8	43
20	1600	2.7	43
21	1680	2.6	44
22	1460	2.6	45
23	1840	2.5	46
24	1920	2.4	46
25	2000	2.3	46
26	2080	2.3	48
27	2160	2.2	48
28	2240	2.2	49
29	2320	2.1	49
30	2400	2.1	50
31	2480	2.1	52
32	2560	2.1	54
33	2640	2.1	55
34	2720	2.1	56
35	2800	2.0	57

From Akiyama, 1993

Table 8-7 Amount of Feed (%) to Be Placed in the Feeding Trays and the Corresponding Consumption Period

Weight (gm)	Amount of Feed in Tray (%)[1]	Consumption Period (hr)
3–4	3.6	2.5
5–8	4.1	2.5
9–12	4.5	2.0
13–19	5.0	2.0
20–28	5.4	1.5
29–34	5.9	1.0
35–40	6.2	1.0

[1]Quantity evenly divided by the number of feeding trays (see Table 7.1)

From Akiyama, 1993

The physiological basis of compensatory growth is poorly understood, but is characterized by both rapid growth and efficient food conversion. These may result from the "filling out" of existing body cells depleted during food deprivation, a process more efficient than the formation of new cells. During short-term starvation, visceral fats and muscle lipids are mobilized as energy sources, and the muscle lipids are replaced with water. When food supplies are restored there is a rapid increase in the weight of the muscle, an increase in the glycogen and lipid content of muscle cells, and a corresponding decrease in the muscle water content (Jobling, 1994).

In view of this apparent efficiency of animals to "catch-up" on body weight after periods of deprivation, and the changes known to occur in muscle composition, trials have been conducted with various types of livestock, including fish, to examine any potential advantages in growth, food conversion or flesh quality to be gained by feed "cycling," that is, following periods of starvation, or restricted feeding with periods of satiation feeding.

A number of studies have been conducted with salmonid fish. The results indicate that short periods of feed deprivation, followed by refeeding to satiation, result in little or no growth loss, when compared to the growth of fish fed daily to satiation (Kindschi, 1988; Miglavs and Jobling, 1989; Thorpe et al., 1990; Quinton and Blake, 1990).

The ability of the fish to compensate for lost growth depends on a number of factors, including species, age, duration of the starvation period, and environmental conditions (Blyth et al., 1992). This makes it difficult to draw conclusions from the existing published data, which has been collected under widely differing experimental conditions. Studies have shown, however, that both rainbow trout (Quinton and Blake, 1990) and channel catfish (Lovell, 1994) can rapidly recover from 3 weeks of restricted feeding, if the period of restricted feeding is followed by satiation feeding at

Table 8-8 Feeding Rate Adjustments Based on the Use of Feed Trays

Average Amount of Unconsumed Feed Remaining on Trays	Adjustment to Feeding Rate
0	Increase 5%
<5	No change
5–10	Decrease 5%
10–25	Decrease 10%
>25	Suspend 2 feed rations; reinitiate at 10% less

From Clifford, 1992

optimum temperatures. In experiments with channel catfish, year-2 fish were fed on restricted rations (once per 3 days) for periods of 3, 6, and 9 weeks, and their growth, food conversion, body fat, muscle diameter, and dress-out yield were compared with data gathered from continuously fed controls over an 18-week trial period (Lovell, 1994). The fish fed restricted rations for 3 weeks weighed the same as the control fish after 3 weeks of full feeding. Fish given restricted rations for 6 weeks weighed 91% of the weight of the controls, and fish fed restricted diets for 9 weeks were 87% of the weight of the controls at the end of the experiment. Carcass analysis values were similar for all groups of fish. Body fat content was similar in all groups, although muscle fiber diameter was significantly smaller in samples taken from fish that had been starved for the first 6 or 9 weeks. These findings, that fish growth rates can be maintained without daily feeding, have implications for feed management. Weekends and holidays make up 31% (114 days) of the year in North America (Kindschi, 1988), and the labor costs associated with feeding and cleaning tanks over this period account for a substantial portion of the farm budget and manpower. While many farmers reduce feeding at weekends, the results of controlled experiments to determine the effects are equivocal. It has been reported that groups of large rainbow trout (>1kg) grew equally when fed to satiation for 6 days and then starved for one day, as groups fed daily to satiation. Reduced growth was recorded, however, in groups fed to satiation for 5 days and then starved for 2 days (Cho, 1992). Commercial-scale experiments with Atlantic salmon in sea cages similarly showed that growth rates were significantly lower in fish fed to satiation for 5 days and then starved for 2 days, than in fish fed to satiation daily (Blyth et al., 1992).

REFERENCES

ABESAMIS, G.C. 1989. Philippine shrimp growout practices. p. 93–101. In D.M. Akiyama, ed. Proceedings of the S.E. Asia Shrimp Farm Management Workshop. American Soybean Association, Singapore.

AKIYAMA, D.M., 1993. Semi-extensive shrimp farm management. Technical Bulletin Vol. AQ 38, pp. 1–20. American Soybean Association, Singapore.

ANDREWS, J.W. and J.W. PAGE. 1975. The effect of frequency of feeding on culture of catfish. Transactions of the American Fisheries Society 104:317–321.

AUSTRENG, E., T. STORBAKKEN, and T. ASGARD. 1987. Growth rate estimates for cultured Atlantic salmon and rainbow trout. Aquaculture 60:157–160.

BLYTH, P.J., G.J. PURSER and C.K. FOSTER. 1992. Can compensatory growth improve the profitability of Atlantic salmon aquaculture? In Proceedings of the Aquaculture Nutrition Workshop, Salamander Bay, April 15–17, 1991, eds. G.L. Allan and W. Dall, pp. 227–231. NSW Fisheries, Brackish Water Fish Culture Research Station, Salamander Bay, Australia.

BOUJARD, T. and J.F. LEATHERLAND. 1992. Demand-feeding behavior and diel patterns of feeding activity in *Oncorhynchus mykiss* held under different photoperiod regimes. Journal of Fish Biology 40:535–544.

BRETT, J.R. and T.D.D. GROVES. 1979. Physiological energetics. In Fish Physiology VIII: Bioenergetics and Growth, eds. W.S. Hoar, D.J. Randall and J.R. Brett, pp. 279–352. Academic Press, New York.

BRETT, J.R., J.E. SHELBORN and C.T. SHOOP. 1969. Growth rate and body composition of fingerling sockeye salmon, *Oncorhynchus nerka,* in relation to temperature and ration size. Journal of the Fisheries Research Board of Canada 26:2363–2394.

BRYANT, P.L. and A.J. MATTY. 1981. Adaptation of carp *(Cyprinus carpio)* larvae to artificial diets. I. Optimum feeding rate and adaptation age for a commercial diet. Aquaculture 23: 275–286.

CARRILLO, M.A., J. PEREZ and S. ZANUY. 1986. Efecto de la hora de injesta y de la naturaleza de la dieta sobre el crecimiento de la lubine *(Dicentrarchus labrax L.)*. Investigación Pesquera 50:83–95.

CHO, C.Y. 1992. Feeding systems for rainbow trout and other salmonids with reference to current estimates of energy and protein requirements. Aquaculture 100:107–123.

CHUA, T.E. and S.K. TENG. 1978. Effects of feeding frequency on the growth of young estuary grouper, *Epinephelus salmoides* Maxwell, culture in floating net cages. Aquaculture 14:31–47.

CLARKE, J.H., W.O. WATANABE, D.H. ERNST and R.I. WICKLUND, 1990. Effect of feeding rate on growth and feed conversion of Florida red tilapia reared in floating marine cages. Journal of the World Aquaculture Society, 21:16–24.

CLIFFORD, H.C. 1992. Marine shrimp pond management. In Proceedings of the Special Session on Shrimp Farming, pp. 301, ed. J. Wyban, pp. 110–137. World Aquaculture Society, Baton Rouge.

CUZON, G., M. HEW and D. COGNIE. 1982. Time lag effect of feeding on growth of juvenile shrimp, *Penaeus japonicus.* Aquaculture 29:33–44.

DEUEL, C.R., D.C. HASKELL, D.C. BROCKWAY and O.R. KINGSBURY. 1952. The New York state fish hatchery feeding chart, 3d ed. Fisheries Research Bulletin, New York 3:61.

FURNELL, D. 1990. The feed squeeze—how to make every pellet pay. Canadian Aquaculture 6:37–39.

HASKELL, D.C., 1955. Trout growth in hatcheries. New York Fish and Game Journal 6(2): 204–237.

HOGENDOORN, H. 1981a. Controlled propagation of the African catfish, *Clarias lazera* (C. & V.) III. Feeding frequency and growth of fry. Aquaculture **21**:233–241.

HOGENDOORN, H. 1981b. Controlled propagation of the African catfish, *Clarias lazera* (C. & V.) IV. Effect of feeding regime in fingerling culture. Aquaculture **24**:123–131.

HUISMAN, E.A. 1976. Food conversion efficiencies at maintenance and production levels for carp, *Cyprinus carpio L.,* and rainbow trout, *Salmo gairdneri R.* Aquaculture **9**:259–273.

JAUNCEY, K., B. ROSS, 1982. A Guide to Tilapia Feeds and Feeding. Institute of Aquaculture, Stirling.

JOBLING, M. 1994. Fish Bioenergetics. Chapman & Hall, London.

JOBLING, J., J.S. Christiansen, E.H. Jørgensen, and A.M. Arnesen. 1993. The application of X-radiography in feeding and growth studies with fish: A summary of experiments conducted on Arctic charr. CRC Reviews Fisheries Science **1**:223–237.

KINDSCHI, G.A. 1988. Effect of intermittent feeding on growth of rainbow trout, *Salmo gairdneri* Richardson. Aquaculture and Fisheries Management **19**:213–215.

KNIGHTS, B. 1983. Food particle-size preferences and feeding behavior in warmwater aquaculture of European eel, *Anguilla anguilla* (L.). Aquaculture **30**:173–190.

LOVELL, T. 1994. Compensatory gain in fish. Aquaculture Magazine **1**:91–93.

MÄKINEN, T. 1993. Effect of feeding schedule on growth of rainbow trout. Aquaculture International **1**:124–136.

MIGLAVS, I. and M. JOBLING. 1989. Effect of feeding regime on food consumption, growth rates and tissue nucleic acids in juvenile Arctic charr, *Salvelinus alpinus,* with particular respect to compensatory growth. Journal of Fish Biology **34**:947–957.

QUINTON, J.C. and R.W. BLAKE. 1990. The effect of feed cycling and ration level on the compensatory growth response in rainbow trout, *Oncorhynchus mykiss.* Journal of Fish Biology **37**:33–41.

REYMOND, H. and J.P. LAGARDÈRE. 1990. Feeding rhythms and food of *Penaeus japonicus* Bate *(Crustacea, Penaeidae)* in salt marsh ponds: role of halophilic entomofauna. Aquaculture **84**:125–143.

ROBERTSON, L., A.L. LAWRENCE and F.L. CASTILLE. 1993. Effect of feeding frequency and feeding time on growth on *Penaeus vannamei* (Boone). Aquaculture and Fisheries Management **24**:1–6.

SEDGEWICK, R.W. 1979. Effect of ration size and feeding frequency on the growth and feed conversion of juvenile *Penaeus merguiensis* De Man. Aquaculture **16**:279–298.

SIRAJ, S.S., Z. KAMARUDDIN, M.K.A. SATAR, and M.S. KAMARUDIN. 1988. Effects of feeding frequency on growth, food conversion and survival of red tilapia *(Oreochromis mossambicus/O. niloticus)* hybrid fry. In The Second International Symposium on Tilapia in Aquaculture, ICLARM Conference Proceedings 15, p. 623, eds. R.S.V. Pullin, T. Bhukaswan, K. Tonguthai, and J.L. Maclean, pp. 383–386. Department of Fisheries, Bangkok, Thailand and ICLARM, Manila, Philippines.

SMITH, I.P., N.B. METCALFE and F.A. HUNTINGFORD, 1995. The effects of food pellet dimensions on feeding responses by Atlantic salmon *(Salmo salar L.)* in a marine net pen. Aquaculture **130**:167–175.

STOREBAKKEN, T., E. AUSTRENG and K. STEENBERG. 1981. A method for determination of feed intake in salmonids using radioactive isotopes. Aquaculture 24:133–142.

STRADEMEYER, L. 1992. Appearance and taste of pellets influence feeding behaviour of Atlantic salmon. In: The Importance of Feeding Behavior for the Efficient Culture of Salmonid Fishes, ed. J.E. Thorpe and F.A. Huntingford, pp. 21–28. World Aquaculture Society, Baton Rouge.

TABACHEK, J.L. 1988. The effect of feed particle size on the growth and feed efficiency of Arctic charr (Salvelinus alpinus). Aquaculture 71:319–330.

TALBOT, C. 1994a. How growth relates to ration size. Fish Farmer January/February:45–46.

TALBOT, C. 1994b. Time to feed. Fish Farmer July/August:49–50.

TALBOT, C. and P.J. HIGGINS. 1983. A radiographic method for feeding studies on fish using metallic iron powder as a marker. Journal of Fish Biology 23:211–220.

THORPE, J.E., C. TALBOT, M.S. MILES and D.S. KEAY. 1990. Control of maturation in cultured Atlantic salmon, Salmo salar, in pumped sea-water tanks, by restricting food intake. Aquaculture 86:315–326.

TUCKER, J.W., 1992. Marine fish nutrition. In: G.L. Allan and W. Dall, eds., Proceedings Aquaculture Nutrition Workshop, Salamander Bay, 15–17 April 1991. NSW Fisheries, Brackish Water Fish Culture Research Station, Salamander Bay, Australia, pp. 25–40.

WANKOWSKI, J.W.J. and J.E. THORPE. 1979. The role of food particle size in the growth of juvenile Atlantic salmon (Salmo salar L.). Journal of Fish Biology 14:351–370.

9
↘↘↘↘↘↘

Performance Measures

INTRODUCTION

The success of feeding operations on fish and shrimp farms is reflected in overall productivity and cost effectiveness. Survival, growth rates, and feed conversion ratios are the main parameters which are routinely determined by fish farmers. They are used to evaluate performance and to plan future feed operations. In addition to these parameters, measures of product quality should also be taken, particularly where these are directly related to issues of feed policy.

Effective feed management must be based on the assessment and interpretation of data gathered by regular sampling of fish or shrimp stocks. Sampling involves weighing or measuring a representative group (subsample) of animals from within populations. These data can then be used to determine the total biomass present, and the changes in weight or length that have occurred since the previous sampling date. The culture species and type of rearing system employed will influence the sampling techniques used. All methods, however, should be carefully designed to avoid bias in the sampling method, and to minimize any stresses exerted on the animals during sampling.

SAMPLING METHODS

Shrimp

In the early stages of shrimp culture operations the growth and survival rates of post-larval shrimp in nursery ponds are generally estimated from observations as they

159

gather on feeding trays. The first true samplings are normally conducted 30–45 days after transfer to production ponds, when the shrimp are large enough (>1 g) to be retained in a cast-net, and then regularly at 7-day intervals until they are harvested (Cruz, 1991). While cast-nets are most commonly used to capture shrimp during sampling, some farmers prefer to use seine-nets (Fig. 9.1) or lift-nets, and may sample post-larval shrimp as early as two weeks after stocking, using a fine-mesh seine net (Samocha and Lawrence, 1992).

Whichever capture method is used, shrimp should be collected from a number of sites within each pond and in sufficient numbers to accommodate the variations in size distribution which are found in all production ponds. Samples should be taken from both deep and shallow regions of each pond, since larger animals often congregate in the deeper areas, while smaller animals often gather along the shallower edges of ponds. Sampling bias may also occur if samples are taken immediately after feeding, since larger animals forage more efficiently and may be present in disproportionately high numbers in feeding areas (Cruz, 1991). Similar biases may occur where animals for sampling are collected from feeding trays.

Ponds used for the grow-out of shrimp range in size from 0.5–40 ha and generally there is greater size variation with increasing pond size. Where this occurs it is recommended that sample size should be made relative to the pond size. Cruz (1991) has devised a cast-net sampling scheme for ponds ranging from 0.4–4.0 ha based on inten-

Fig. 9.1   Sampling farmed shrimp using a small-mesh seine net.

Table 9-1 Recommended Cast-net Sampling Scheme for Different Pond Areas (Effective Cast-net Area is Assumed at 4m²)

Pond Area (m²)	Typical Stocking/m²	Sampling Area %	Sampling Area m²	Cast-net Throws	Approx. No. of Animals
5,000	15–20	0.3	16	4	225–300
10,000	10–15	0.3	32	8	300–450
20,000	5–10	0.3	60	15	300–600
40,000	2.5–5	0.3	120	30	300–600

From Cruz, 1991

sive and semi-intensive pond culture practices for *Penaeus monodon* in the Philippines (Table 9-1). Captured shrimp should be separated from debris and mud and temporarily held in containers of clean, oxygenated water before measuring.

Fish

Fish should be starved for up to 24 h before they are sampled in order to reduce the stressful effects of netting and handling. The commonest technique for sampling fish involves temporarily crowding the fish into one section of the pond, tank, or cage, either by lowering the water level or using a net or screen. Samples of fish are then netted out for weighing and counting. 1% or more of the estimated population (≥100 animals per 10,000) should be sampled.

For the collection of weight and length data from individual fish, a minimum number of 30 fish (ideally 50) should be taken randomly from each sample and held in containers of well-oxygenated water.

MEASUREMENTS

Mean Weights

Plotting changes in mean (average) weights of fish and shrimp populations throughout production cycles is the common practice for most fish farmers. The total weights of sampled fish or shrimp are determined by batch-weighing, usually in containers of water on a tared balance. The mean weight of the sample is then determined by dividing the total weight by the number of animals. It is not normally necessary to anaesthetize fish prior to batch weighing since the operation can be performed relatively quickly. There are no shortcuts to accurate sampling however. Sampling, measuring, and recording equipment should be assembled, and a work schedule organized before any animals are collected.

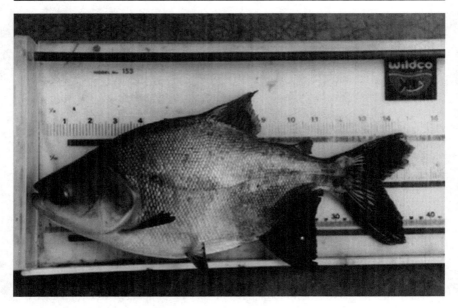

Fig. 9.2   Fish measuring board. Using a graduated fish measuring board to determine the fork length of a tambaqui, *Colossoma macropomum*.

Length and Weight Data

Length and weight data gathered from individual fish and shrimp can be used to check for any sampling bias, to estimate the size diversity within the population, and to calculate condition indexes. Fish should be anaesthetized prior to handling for length measurement and individual weighing. A graduated measuring board with an end stop is generally used for length measurements of fish (Fig. 9.2). The board can be rested directly on the balance so that weight (±0.1 g) and length data (±1 mm) can be recorded simultaneously from each fish. Other methods for measuring fish length include the use of calipers, tape measures, or boards with movable crosshairs, which are moved along a scale above the fish (Ricker, 1979). The common measure used in aquaculture is the fork length (Fig. 9.3), sometimes referred to as the median length. Measurements of total length may however be more useful for those species such as tilapias, which lack a distinct tail fork. Whichever measure is taken, the important issue is to adapt a fish handling and measuring procedure that is consistent and repeatable (Fig. 9.4).

As fish grow, their changes in weight are relatively greater than their changes in length. This results from the approximately cubic relationship between fish length and weight. This generally means that determining changes in the weight of fish under practical farming conditions, and over short growth periods, are more useful than length measurements, particularly since farmers sell their products by weight.

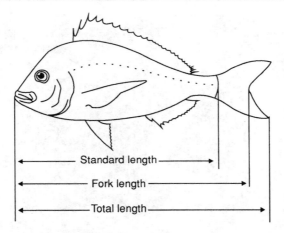

Fig. 9.3 Length measurements for fish.

Length measurements of shrimp and prawns may be taken from various points. Most common are measures of the postorbital–carapace length, measured from the posterior middorsal region to the posteriormost part of the orbit, total carapace length, or the total extended length, from the tip of the rostrum to the tip of the telson. Where the rostrum is frequently damaged the former measure may be the more accurate (Fig. 9.5). Both weight and length measurements are used for sampling growth in shrimp. Variable amounts of water held in the carapaces of shrimp may affect the accuracy of weight measurements (Fig. 9.6).

Graphs relating length to weight are useful in growth determinations of both fish and shrimp. Where it is easier, for whatever reason, to determine only one set of parameters, the other can be estimated from existing graphs. For accuracy, length–weight graphs should be based on data collected from animals of similar genetic makeup, and which were cultured under similar conditions (Fig. 9.7).

Stressful Effects of Handling

Handling fish during routine sampling activities results in various levels of adaptive stress responses. The primary responses are physiological and involve the release of catecholamines and corticosteroids. These compounds released from the autonomic nervous system and the kidneys, respectively, induce secondary responses in metabolism, osmoregulation, and energetic transformations. The visible effects of these temporary changes may be seen in loss of appetite, abnormal feeding behavior, regurgitation of food, and defecation. After handling, these responses progressively return to normal levels. Reductions in the levels of stomach contents, and lowering of circulating nutri-

Fig. 9.4 Sampling farm stocks. Taking simultaneous weight and length measurements of Atlantic salmon.

ents, are typically compensated by a temporary increase in food intake. Depending on the species, and the frequency and intensity of the stresses involved in sampling, it may take from just several hours up to 7 days for fish to resume normal feeding. Generally, however, normal feeding is resumed after 1–2 days. A detailed study of the effects of sampling on farmed gilthead sea bream, while supporting these general trends seen in fish, also showed that suppression of feeding after handling and weighing was less marked in smaller than in larger fish (Kentouri et al., 1994).

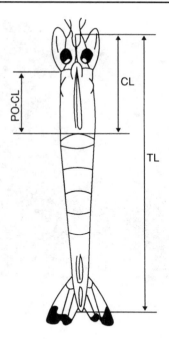

Fig. 9.5 Length measurements for shrimp. PO-CL = Post-orbital carapace length; CL = Carapace length; TL = Total length.

GROWTH PATTERNS

Fish and shrimp do not grow continuously throughout their lives. In temperate climates the pattern of growth is closely linked to seasonal changes in water temperature and food availability. In tropical and subtropical regions, other factors, such as rainfall, or salinity, may also cause seasonal checks in growth. Partitioning of energy varies with the stage of development of fish and shrimp. Larvae and juveniles use energy primarily for growth, and secondarily for maintenance, while adult and sexually mature animals expend much of their energy intake on reproduction and maintenance. These changes are reflected in a generalized growth curve (Fig. 9.8). The sigmoid or S-shaped curve represents several phases of growth (Hopkins, 1992). The early rapid growth of larvae and fry follows an exponential curve (points A–C) while juvenile, subadult growth follows a linear relationship (points B–D). Growth then slows as energy is channeled into reproductive activities and maximum size is approached (points C–E). The growth of shrimp, and other crustaceans, follows discrete, stepwise progressions, which correspond to molting periods, and result from the rapid addition of water to newly functional tissues. The underlying pattern of growth, in terms of tissue elaboration, is similar to that of all fish, however.

Fig. 9.6   Using calipers to measure the postorbital–carapace length of a penaeid shrimp.

GROWTH RECORDING

Measurements of farmed fish and shrimp may be based on absolute changes in length or weight (absolute growth) or changes in length or weight relative to the size of the fish (relative growth). Measurements of growth expressed in terms of time intervals (days, weeks, or months) constitute growth rates. A data set of weight–length measurements of tambaqui, *Colossoma macropomum*, collected from growth trials in Brazil (Table 9-2) is used here to illustrate the various measures.

i. Absolute growth in weight from Day 0 to Day 192

$$= W_2 - W_1 \qquad\qquad (9.1)$$
$$= 633 - 12 = 621 \text{ g}$$

where W_1 is the initial mean weight and W_2 the final mean weight.

ii. Absolute growth rate

$$= \frac{W_2 - W_1}{t} \qquad\qquad (9.2)$$
$$= \frac{633 - 12}{192}$$
$$= 3.2 \text{ g/day}$$

where t is the length of the growth period in days

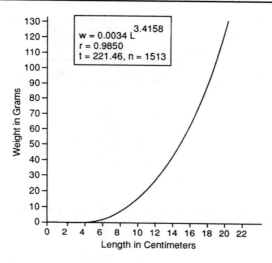

$$w = 0.0034 \, L^{3.4158}$$
$$r = 0.9850$$
$$t = 221.46, \; n = 1513$$

Fig. 9.7 Length–weight data for the giant freshwater prawn. From Menasveta and Piyati-ratitivokul, 1982.

The absolute growth rate is a more useful measure of growth than the absolute growth and is widely used by fish farmers as a practical measure to monitor and compare the performance of fish stocks. The method used to derive absolute growth rates is based on the assumptions that the gains in weight against time follow a linear relationship, and that the absolute growth rate is the same regardless of the size of the fish

Fig. 9.8 A sigmoid growth curve.

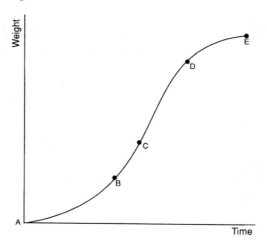

Table 9-2   Data Set for Growth of Tambaqui, *Colossoma macropomum,* from Pond Culture Trials in Northeast Brazil (n=2500)

Day	Length (cm)	Weight (g)
0	8.9	12
31	18.3	113
66	25.4	293
94	28.1	388
128	30.2	507
156	31.7	551
192	35.0	633

Source: Bezerra (personal communication)

These assumptions may or may not hold true over the sampling period. For many species growth is approximately linear over the typical rearing cycle of a fish farm, where fish are grown from fingerling size and are harvested prior to the onset of sexual maturity.

Absolute growth rates are often used, with analysis of variance (ANOVA) procedures, to evaluate various treatments of farmed fish or shrimp populations, such as comparing the yields from groups of animals fed on different diets. The method is applicable where all the test fish or shrimp have approximately the same starting weight, and where the growth trials are conducted over the same period of time, and under identical environmental conditions.

iii. Instantaneous growth rate

The instantaneous growth rate (G) is a measure of the exponential growth typical of young fish. It describes growth at a particular instant of time rather than over a long time period.

$$G = \frac{L_n(W_2) - L_n(W_1)}{t} \qquad (9.3)$$

where $L_n(W_2)$ is the natural logarithm of the weight at time t, and $L_n(W_1)$ is the natural logarithm of the initial weight.

It is common practice with fish farm data to multiply G by 100 and to express the result as specific growth rate (SGR) in %/day. From the data from Table 9-2:

$$SGR = \frac{L_n(W_2) - L_n(W_1)}{t} \times 100 \tag{9.4}$$

$$= \frac{L_n(633) - L_n(12)}{192} \times 100$$

$$= \frac{6.45 - 2.48}{192} \times 100$$

$$= 2.06\%/day$$

The specific growth rate is a widely used measure for recording the growth performance of fish and shrimp. Specific growth rates are not subject to the same restrictions on initial size or time period as absolute growth rates, when used to evaluate and compare growth data.

Length–Weight Relationships

For most species, growth in weight (W) increases as an exponent of length. The relationships of fish weights and lengths are expressed by the power functions:

$$W = aL^b \text{ (used for the prediction of W from L)} \tag{9.5}$$
$$L = aW^b \text{ (used for the prediction of L from W)} \tag{9.6}$$

These equations can be transformed into linear form using the following equations:

$$\log_{10}(W) = \log_{10}(a) + b\log_{10}(L) \tag{9.7}$$
$$\log_{10}(L) = \log_{10}(a) + b\log_{10}(W) \tag{9.8}$$

In Equations 9.6 and 9.8 the value of the exponent b is usually close to 3 since growth in length takes place in a single dimension, while weight (volumetric growth) occurs in three dimensions. The value of b in equations 9.5 and 9.7 will generally lie between 0.3 and 0.4. In establishing length–weight relationships, large sample sizes should be used, and the correlation coefficient associated with the regression should be very high, exceeding 0.9. Fig. 9.7 illustrates the length and weight relationship of the giant freshwater prawn, *Macrobrachium rosenbergii*. Measurements from 1513 animals were used to establish the relationship shown (Menasveta and Piyatiratitivokul, 1982).

GROWTH PREDICTION

Many microcomputer models are available to fish farmers as aids to management and planning. Spreadsheet programs can be used to model stocking and harvesting schedules, growth, and feed factors, and to generate cash flow information, and profit and loss accounts. Software programs are available from commercial suppliers, extension agencies, and some feed manufacturing companies.

The ability to predict the growth rate of fish and shrimp is an important management technique. Software programs that predict growth, based on food consumption rates and assumed food conversion values, are widely used. For practical purposes the growth of fish under a given set of conditions can be calculated as a function of the amount of food consumed and the food conversion ratio. Growth can be calculated on a daily or weekly basis using simple equations. For example:

$$W_1 = W_0 + \frac{(W_0 \times R)}{FCR} \tag{9.9}$$

where:

W_1 = Weight at end of week 1
W_0 = Initial weight
R = Total rations fed during the week
FCR = Food conversion ratio

The weight gain can be calculated on a weekly additive basis. At the end of week 2 the weight of the fish (W_2) will be given as:

$$W_2 = W_1 + \frac{(W_1 \times R)}{FCR}$$

This calculation is continued forward in time until target weights or grading intervals are reached. Software programs are available that accumulate data for numbers and sizes of fish and which offer daily and weekly updating. The steering parameters of FCR and feeding rate are adjustable and are set for individual operations. The two major factors affecting fish growth are water temperature and fish size. These must be taken into account when using simple growth prediction models. Most available software is based on daily feed amounts derived from feeding tables, which take into account these variables. The calculations as illustrated, are based on the average weight of fish populations, and do not take into account the weight distribution of the population. A limitation in the use of growth models of this type is the need to assume values for food conversion ratios.

CONDITION FACTORS

The curvilinear relationships between weight (W) and length (L) can be used to determine condition factors for fish. In fisheries biology, the condition factor is used to measure the variation from the expected weight for length of individual fish, or groups of individuals, as an indication of differences in fatness, changes in nutritional status, environmental effects, sexual modes, and body shape (LeCren, 1951). Each fish species has a characteristic range of condition factors, which reflects their body conformation.

Species such as rainbow trout and Atlantic salmon are slender-bodied species and have lower condition factors than thick-bodied species such as common carp.

The condition factor is generally derived from the equation:

$$k = \frac{Weight\ (g)}{Length\ (cm)^3} \times 100 \qquad (9.10)$$

where k = Condition factor

Genetic strains of some species may have characteristic k values which differ from those of other strains. This is seen for example among the many strains of common carp reared throughout the world, which have been variously selected for the flesh yields associated with a deep-body conformation. Significant differences also occur in fish sampled at different stages, or phases, of production cycles. For example, the condition factors of large rainbow trout grown in the sea are much higher than those recorded from the portion-size trout typically reared on freshwater farms. Table 9-3 lists some examples of condition factors.

For individual species, or genetic strain variations, condition factors may be accounted for by the fullness of the digestive tract, degree of sexual maturity, or the dietary condition (Steffens, 1989). In farmed fish, condition indexes may be used to confirm visual inspections as to whether the fish have a typical conformation or are too fat or too thin. The results obtained may predicate a change of feeding level or a switch to a feed type with a different nutrient density.

SURVIVAL

Accurate recording, or estimation, of survival levels of fish and shrimp in culture systems is of fundamental importance in calculating feed rations, and estimating the standing biomass. A reasonably accurate knowledge of the numbers of animals in each holding system is also vital to production and harvest planning. The practices adopted for monitoring survival vary considerably depending on the species and rearing system.

Table 9-3   Condition Factors

Species	Condition Factor
Rainbow trout	1.3–1.6
Atlantic salmon	1.0–1.2
Channel catfish	1.0
Common carp	2.0–2.5

Estimating Survival of Fish

In intensive fish culture an accurate count of juveniles is generally possible at the time that fish are transferred from the hatchery to ponds, raceways, tanks, or cages, for the main grow-out stage. Further counts may be made when the stocks are later graded. Fish may be individually counted, by hand or automatic counter (Fig. 9.9), into holding systems, or more commonly, given the large numbers of small animals involved, a gravimetric method is used. This involves determining the numbers of fish per unit of weight (n/kg), and then weighing and transferring the required biomass. The accuracy of the weighing will determine the overall numbers transferred.

After transfer, survival estimates are based on the accurate recording of daily mortalities, and subtraction of their weights from the standing biomass. The opportunities to observe and remove dead fish varies with the nature of the holding system. In tanks and raceways with good water visibility, accurate counts and removals can generally be made on a daily basis, as part of routine husbandry. Mortalities in cages and ponds are more difficult to monitor. Dead fish are not usually visible at the bottom of deep cages and regular inspections must be made either by divers or by lifting the net. In pond systems, dead fish that sink to the bottom often pass unrecorded, and farmers may have to estimate an average mortality, based on previous stocking and harvesting data.

Whichever method is used to estimate survival it should be borne in mind that much of the overfeeding which occurs in intensive aquaculture arises because farmers are overestimating the standing biomass. It is not uncommon for farmers to incorrectly estimate biomass within ±25%, or more, of the actual amount, figures that are often not revealed until the fish are harvested, and resulting losses associated with the husbandry and feeding of nonexistent fish calculated. Such inaccurate biomass estimates seriously compromise feed and stock management. Farmers should ideally know the biomass of their stocks to within ± 5%.

Estimating Survival of Shrimp

Estimating shrimp survival in pond systems poses considerable problems, since once stocked for final grow-out they are inaccessible until harvest when ponds are drained down. Three methods are used to estimate standing biomass: by back-calculation from estimates of food demand, by using cast-nets to sample densities, or the use of standard survival data (Cruz, 1991). The most reliable methods are estimations of survival based on food demand, where:

$$\text{Standing biomass} = \text{Daily feed consumption(kg)/Feeding rate} \qquad (9.11)$$

and

$$\text{Survival (\%)} = (\text{Standing biomass(kg)/Average weight(kg)/Total no. of shrimp stocked}) \times 100 \qquad (9.12)$$

Fig. 9.9 An automatic fish counter. Counting Atlantic salmon smolts into net-pens during transfer.

The accuracy of using feed demand to determine survival rests on the reliability of estimates of feed demand, feeding rates, and the numbers of shrimp initially stocked (Cruz, 1991). This necessitates careful use of appropriate numbers of feeding trays.

It is common practice for shrimp farmers to base survival estimates on data collected from previous production cycles. Where records show, for example, that 60% of shrimp survived from stocking to harvesting, the assumption is made that, barring catastrophic loss, similar survival levels will be achieved. Over the length of the production cycle, losses are computed based on the overall 40% loss of the total number of animals initially stocked. Losses are typically heaviest in the early stages of the production cycle, following the stresses associated with transfer, and while the animals are most vulnerable to predation. This bias should be taken into account when estimating the weekly or monthly losses. Some shrimp farmers use survival nets, small netted enclosures containing known numbers of post larvae, to estimate the survival of shrimp during transfer to nursery ponds.

FOOD CONVERSION

Measures of weight gain, combined with measures of the amount of food used to produce that weight gain, are used to determine food conversion ratios (FCR) according to the formula:

$$FCR = Food fed/Weight gain \qquad (9.13)$$

For example, a population of 10,000 tilapia have been fed 4% of their body weight per day for a total of 20 days. The individual mean weight of the tilapia was 15 g at the start of the feeding period, and 23 g when the fish were sampled after 20 days.

$$
\begin{aligned}
\text{Feed fed} &= 10,000 \times 15 \times 20 \times .04/1000 \text{ kg} \\
&= 120 \text{ kg} \\
\text{Weight gain} &= (23 - 15) \times 10,000/1000 \text{ kg} \\
&= 80 \text{ kg} \\
\text{FCR} &= \text{Food fed/weight gain} \\
&= 120/80 \\
&= 1.5
\end{aligned}
$$

This should be written completely as the ratio 1.5:1, or 1.5 kg food fed:1 kg of weight gain, but is most commonly abbreviated to 1.5.

Food conversion efficiency (FCE), or food efficiency, is the reciprocal of the FCR converted to a percentage value:

$$
\begin{aligned}
\text{FCE (\%)} &= \text{Weight gain/food consumed} \times 100 \qquad (9.14) \\
&= 80/120 \times 100 \\
&= 66.6\%
\end{aligned}
$$

In intensive aquaculture food conversion values are generally calculated based on the conversion of dry feed to wet fish flesh. Under these terms, they are used to describe a practical economic relationship as opposed to describing the true biological conversion. An FCR of 1.2:1, based on the conversion of dry feed to wet fish flesh, may describe a true conversion value of 3–4:1, based on the amount of raw fish that was converted to fish meal and then blended with other ingredients, to manufacture a dry diet.

Where food conversion values are used to compare the performance of feeds with different moisture values (dry vs. moist feeds), moisture values should be corrected when calculating the weights of food fed. Dry commercial feeds may contain between 6–10% of moisture, while moist feeds typically contain between 28–35% moisture.

Calculations of food conversion ratios may be made periodically throughout the growth cycle, whenever the fish or shrimp stocks are sampled for growth, or calculated on the basis of food use during the complete production cycle. This is sometimes termed the cumulative FCR, and is an important measure for use in the determination of profitability (Chapter 10).

Periodic FCRs are used by many fish farmers as indicators of the effectiveness of both feeding practices and overall husbandry. Many factors can influence food conversion values. Poor or fluctuating FCRs may reflect problems with feed or feeding methods, or may be indicators of wider problems such as the occurrence of disease in stocks, or deteriorating water quality.

ASSESSING QUALITY

Issues of product quality in farmed fish and shrimp are difficult to define since they vary from one market to another. Body size, color and external appearance are important factors. Large animals generally demand a higher price per kilogram, and buyers generally select on the basis of a range of factors. These are shown in Table 9-4.

The quality of the flesh of cultured fish and shrimp is a vital factor that is influenced directly by the nature and balance of feed ingredients (Cowey, 1991). Both dietary proteins and lipids play major roles in the sensory qualities of fish, where taste, odor, and texture are major determinants of the acceptability of products to consumers. In addition to diet, a range of genetic and environmental factors may also affect the flesh composition of fish. These are illustrated in Fig. 9.10.

Table 9-4 Important Factors in the Assessment of Fish Quality

Quality Parameters	Measurements
Processing loss	% weight
Yield	% weight
Fillet color (salmonids)	Minolta chromameter, Roche color card, chemical analysis
Texture	Taste panel, texture meter
Chemical composition	Chemical analysis
Sensory quality	Taste panel

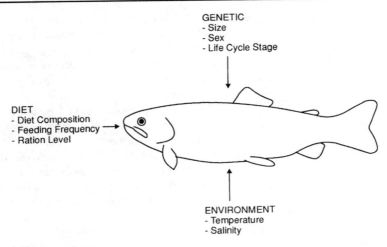

GENETIC
- Size
- Sex
- Life Cycle Stage

DIET
- Diet Composition
- Feeding Frequency
- Ration Level

ENVIRONMENT
- Temperature
- Salinity

Fig. 9.10   Factors affecting the flesh composition of farmed fish. After Shearer, 1994.

Lipids

The quality and quantity of food fed to fish affects their proximate composition in a number of ways. Energy greater than maintenance requirement is stored in the body as lipids. The distribution of this stored lipid in the body varies for different species. It may be stored within the muscle, around the visceral organs, or in the liver. Fat stored in the liver and around the body cavity is often lost as processing waste, and represents an economic loss to either the producer or processor. The amount of fat stored in the muscle of fish affects its texture, taste, and storage properties. Slightly higher than normal fat levels may be sought in fish such as salmonids and eels, which are prepared for smoking. In salmonids up to 12% fat content is typical for smoked product.

The fatty acid profile of the diet is known to influence the fatty acid profile of deposited lipids, which in turn affect the organoleptic properties of fish. The levels of stored lipids are inversely related to the water content, and vary more widely among groups of fish than does the protein level, which is relatively stable for any particular species.

Fat levels in fish flesh can be measured using chemical analysis, or by the use of meters, such as the Torry fat meter. The Torry fat meter, while expensive, is nondestructive and gives rapid results. Some feed manufacturers lend meters to their customers to monitor fat levels in farmed fish. Fat levels must be monitored throughout the growth cycle where specific fat levels are sought for particular markets or secondary processing.

Protein

In contrast to the patterns of uptake of dietary lipids by fish, there are conflicting reports as to the effects of dietary protein levels on the protein content of fish flesh, and no evidence that whole body amino acid profiles can be influenced by diet (Shearer, 1994). A strong correlation exists, however, between body protein levels and body weight. Dietary protein levels and amino acid content may significantly influence muscle fiber recruitment, affecting growth, texture, and muscle quality.

Color

The flesh color of salmonids is a vital quality factor, which must be carefully matched to specific market requirements. Pigmenting fish is an expensive process and should be carefully monitored during the production cycle. Pigmentation levels can be determined against color charts (Fig. 9.11), or by using the Minolta chromameter. Uptake of pigment often varies among fish of the same batch, and sample sizes should be sufficiently large to give a representative value. Producers using visual inspection and color charts should establish a consistent routine and viewing conditions, since perceived color is dependent on the source of illumination and the color of the surroundings. Perceived color may also be influenced by the amount of fat, or mucosal layers present (Shearer, 1994). The Minolta Chromameter provides a consistent quantitative measure of flesh color. Some feed companies provide loans of these meters to their customers for routine monitoring of pigment levels.

Fig. 9.11 Fish pigmentation. Using a color card to estimate the flesh pigmentation levels of Atlantic salmon. Photo courtesy of Sid Hann.

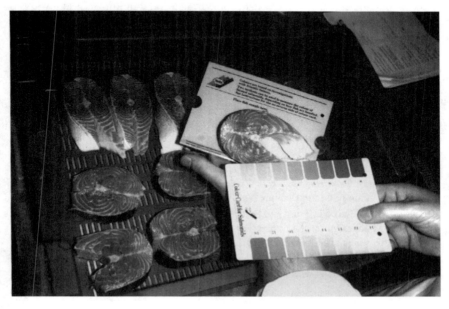

Sensory Quality

Farmed fish and shrimp approaching harvest should be tasted by the same panel of three farm staff at least weekly. They should score taste, texture, and taints, using conventional ranking and profiling methods (Springate, 1993; Stone and Sidel, 1993).

DIETS AND FLESH QUALITY

The selection of particular feed types and ration sizes may be used to modify the flesh quality of farmed fish. In some sectors of the industry, farmers have access to wide ranges of diets. For example, trout diets available in Europe have oil levels ranging from 10–27% of the total diet, and protein levels ranging from 40–48% of the total diet (Springate, 1993). Feeds are also available with different amounts of pigment. These diets are presently used to meet specific quality criteria for gross lipid content and color.

Feeds and feeding practices are not sufficiently developed at present to permit modification of the fatty acid contents of farmed fish. The potential exists, however, since it is known that the total lipid content of salmonids can be influenced by ration size (Johansson et al., 1995), that the profile of stored lipid reflects that of the diet (Sargent et al., 1989), and that the dietary fatty acid content can influence taste (Waagbo et al., 1993). The beneficial effects of lipids from marine fish on human health are well documented. n–3 fatty acids have been attributed roles in the prevention of arteriosclerosis and cardiovascular disease, although the precise mechanisms and extent of these effects are still under debate (Schmidt and Dyerberg, 1994). It can be anticipated that growing public awareness of the benefits of marine fish oils in human health will promote future demand for n–3 enhanced salmonids.

REFERENCES

COWEY, C.B. 1991. Some effects of nutrition on flesh quality of cultured fish. In Fish Nutrition in Practice, eds. S.J. Kaushik and P. Luquet, pp. 227–236. INRA, Paris.

CRUZ, P.S. 1991. Shrimp Feeding Management—Principles and Practices. Kabukiran Enterprises, Inc., Davao City.

HOPKINS, K.D. 1992. Reporting fish growth: A review of basics. Journal of the World Aquaculture Society 23:173–179.

JOHANSSON, L., A. KIESSLING, T. ÅSGÅRD, and L. BERGLUND. 1995. Effects of ration level in rainbow trout, Oncorhynchus mykiss (Walbaum), on sensory characteristics, lipid content and fatty acid composition. Aquaculture Nutrition 1:59–66.

KENTOURI, M., L. LEÓN, L. TORT, and P. DIVANACH. 1994. Experimental methodology in aquaculture: Modification of the feeding rate of the gilthead sea bream Sparas aurata at a self-feeder after weighing. Aquaculture 119:191–200.

LECREN, E.D., 1951. The length-weight relationship and seasonal cycle in gonad weight and condition in the perch (Perca fluviatilis). Journal of Animal Ecology 16:188–204.

MENASVETA, P. and S. PIYATIRATITIVOKUL. 1982. Effects of different culture systems on growth, survival, and production of the giant freshwater prawn (*Macrobrachium rosenbergii*, De Man). In Giant Prawn Farming, ed. M.B. New, Developments in Aquaculture and Fisheries Science, p. 10. Elsevier, Amsterdam.

RICKER, W.E. 1979. Growth rates and models. In Fish Physiology VIII. Bioenergetics and Growth, eds. W.S. Hoar, D.J. Randall and J.R. Brett, pp. 678–743. Academic Press, New York.

SAMOCHA, T.M. and A.L. LAWRENCE. 1992. Shrimp nursery systems and management. In Proceedings of the Special Session on Shrimp Farming, ed. J. Wyban, pp. 87–105. World Aquaculture Society, Baton Rouge.

SARGENT, J., R.J. HENDERSON and D.R. TOCHER. 1989. The lipids. In Fish Nutrition, ed. J.E. Halver, pp. 154–209. Academic Press, New York.

SCHMIDT, E.B. and J. DYERBERG. 1994. Omega–3 fatty acids—current status in cardiovascular medicine. Drugs, 47:405–424.

SHEARER, K.D. 1994. Factors affecting the proximate composition of cultured fish with emphasis on salmonids. Aquaculture 119:63–88.

SPRINGATE, J. 1993. Farmed fish are what they eat. Fish Farmer **Nov./Dec.**:19.

STEFFERS, W. 1989. Principles of Fish Nutrition. Halsted Press, New York.

STONE, H. and J.L. SIDEL. 1993. Sensory Evaluation Practices. Academic Press Inc., New York.

WAAGBO, R., K. SANDNES, O. TORRISSEN, A. SANDVIN, and O. LIO. 1993. Chemical and sensory evaluation of fillets from Atlantic salmon fed three levels of N-3 polyunsaturated fatty acids at two levels of vitamin E. Food Chemistry 46:361–366.

10

****

Cost Factors

INTRODUCTION

Feed costs in intensive aquaculture typically range from 30–60% of variable operating costs. In general, the simpler the production technology the higher will be the proportion of the feed costs within the operating budget of the farm. Effective feed management involves the control of these operating costs, while at the same time maximizing farm output levels and maintaining product quality.

Many interrelated factors must be taken into account when the feed costs for any particular operation are examined. The focus of these issues differs significantly among the various sectors of the industry. Farmers in established sectors of the industry have access to a wide choice of feed types, produced by competing manufacturers. In emerging sectors, farmers may operate with a restricted choice of feed types, and considerably less support from their feed supply companies in tailoring feed supplies to their specific needs. Farmers in different regions, and sectors of the industry, also operate within widely varying regulatory frameworks for environmental protection. The extent of these controls will have an impact on the selection of feeds and feeding practices. All farmers, however, must meet growing consumer demands with regard to product type and quality, and issues of safety. The interrelationships of the various factors influencing feed management policy are illustrated in Fig. 10.1.

The potential to reduce feed-related costs exist in several areas. These include:

- The use of appropriate feed types.
- The determination of the most cost-effective ration sizes.
- The reduction of food wastage

181

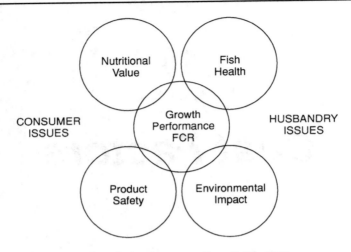

Fig. 10.1   Issues in aquaculture feed management. From Smith, 1990.

Each of the above is dependent on the implementation of accurate data collection systems. All determinations of the costs and benefits relating to feed management issues are dependent on the completeness and accuracy of data gathered on the farm. The absence of accurate records of biomass, feed use, food conversion values, and growth, will compromise all feed management decisions. Data recording should be computerized and data entries made daily. There are numerous ready-made data management programs designed for aquaculture applications. These reduce the chore of record keeping, and permit the rapid manipulation of data. Details of such programs can be obtained from trade journals, training institutions and extension agencies, and the major food supply companies.

SELECTING FEEDS

Where farmers have access to feeds from different sources, selection is generally made on the basis of cost and perceived quality. Further considerations are given to factors such as credit arrangements and the various extension services offered by feed supply companies. In addition to advice concerning feedstuffs, these may include health diagnostic services for customers, loans of equipment, and provision of farm management software. These additional services vary greatly among feed supply companies, but can represent significant benefits for farmers.

In general there is little comparative performance data concerning commercial feed types available to farmers. Some government and farmer-cooperative organizations operate experimental farms, and are able to conduct some trials. Funding for this type of activity is generally limited, however, and distribution of results may generate unwelcome controversy (Maguire, 1992). This situation probably reflects both the stage of maturity of the industry, and the substantial costs involved in conducting farm-scale feeding trials. As the industry continues to develop and grow, fish and shrimp farmers can anticipate a level of independent advisory service, and availability of product information, which compares with that currently available to other livestock farmers.

While individual aquaculture feed manufacturing companies may report trial results for their own products, the practical details of the trial conditions are often sparse, and generally preclude any useful comparisons between competing products. In practice, farmers wishing to evaluate a new feed type need to gather their own information, under the specific operating conditions of their own site. Since few farms have the surplus holding capacity to operate controlled experimental feeding trials, comparisons are normally made on the basis of introducing a new food product into the farm production cycle, and comparing performance of the new product with that of previously used feeds. Comparative data on growth rates and feed conversion ratios can then be linked to the relative costs of the feeds as a basis for future decision making.

The commercial development of aquaculture feeds has progressed rapidly in recent years, extending the choice of feed types available in many sectors of the industry. Farmers must make choices between purchasing finished feeds or making feeds on the

Fig. 10.2   The effects of changes in unit feed price and food conversion ratios on food costs per kg of fish produced. After Crampton and Jackson (1981).

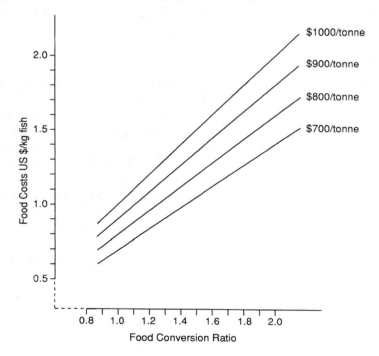

farm, using moist or dry feeds, extruded or steam-pelleted feeds, feeds with different energy and/or protein levels, and the various specialty feeds with additional ingredients such as pigments and attractants. Some choices may be relatively straightforward. Environmental regulations may predicate the use of dry feeds, while farmers seeking floating feeds will need to purchase extruded feeds. Other choices may be more difficult to make. Selecting feeds with different energy levels, whether to buy extruded or less costly pelleted feeds, whether to purchase food in bulk, all represent decisions that must be made on a cost–benefit basis.

Unit Feed Costs

Since feed costs differ, the benefits of more expensive feeds must be evident in terms of improved growth, feed conversion ratios, or product quality. Figure 10.2 illustrates the effects of increased FCR and food price on the food cost per unit of fish produced. From the graph it can be seen that a switch to a more expensive diet must be met by a reduction in the food conversion ratio, if feed costs per unit of production are not to increase. From the Figure it is also evident that small changes in the food conversion ratio have a significant impact on overall feed costs.

Similar trends are seen when alternative feed conversion ratios are analyzed and compared with different feed prices through their impacts on annual rates of return on investment, or their long-term impacts on internal rates of return (Easley, 1977). The impacts of improved FCRs on rates of return generally exceed the effects of changes within the normal range of unit feed prices.

Partial Budgeting

A partial budget can be used to evaluate the costs and benefits of proposed changes in feeding policy. It may eliminate the need to complete a total farm budget where limited changes are being planned in farm operations. These may include plans to intensify inputs, to change technology, or to increase efficiency (Jolly and Clonts, 1993). Partial budgeting involves the identification of all available alternatives, and a determination of their costs and potential benefits. Several steps are involved (Shang, 1990):

- Identify and quantify the gains resulting from the proposed change. These may be in additional gross income, or reduction of costs associated with the change.
- Identify and quantify all of the costs associated with the proposed change. These may be additional costs or reduced income.
- The economic viability of the change can then be evaluated as total gains minus total costs.

where

Profit or loss = (Added income + Reduced costs) − (Added costs + Reduced income).

Table 10-1a   A Theoretical Comparison of Feed Costs Resulting from the Use of Two Feeds, Feed A, and Feed B, Where Feed A Has a Higher Nutrient Density

	Feed Cost/Tonne $US	FCR	Feed Cost/Tonne of Fish Produced $US
Feed A	1050	1.2	1260
Feed B	950	1.4	1330

Table 10-1b   Partial Budget Analysis for Proposed Switch from Food B to Food A, Based on the Benefits and Costs Per Tonne of Fish Produced

Benefits	*Costs*
Added income:	Added costs:
$120 (sales value)	$ 20 (delivery)
	15 (additional storage)
Reduced costs:	Reduced income:
$ 70 (feed)	
5 (bulk purchase)	
20 (veterinary services)	
10 (credit)	
225	35

Profit (Benefits − Costs) = $190 per tonne of fish produced

In selecting feeds it is generally necessary to take into account a wide range of factors, some of which may be difficult to quantify, but need to be considered in the decision-making process. There is, of course, a level of unpredictability in budgeting since production levels on farms vary from year to year, and the purchase prices of feed, or the values of sales may change. Hence, in partial budgeting, best estimates must be applied where costs cannot be accurately predicted.

The example outlined in Table 10-1a illustrates the trade-off between benefits and costs resulting from the use of two feeds, a high-energy feed (A) and a conventional feed (B), which both have the same dry content. The high-energy feed is the more expensive option, but yields superior growth. On the basis of this comparison, Food A would be the food of choice. The benefits of a reduced FCR (diet A) outweigh the benefits of a lower food cost (diet B), indicating that a switch from diet B to diet A would result in food cost savings of $70 per tonne of fish produced.

In Table 10-1b, a wider partial budget analysis is applied to this example. Several additional benefits accrue from the switch from food B to food A, which are supplied by different companies. These include improved credit terms, provision of veterinary services, and reduced bulk purchase costs. There is also an improved quality score, and higher sales value for fish fed on diet A. When these benefits are set against additional delivery and storage costs for food A, the profits resulting from a switch of foods, and supplier, are $190 per tonne of fish produced.

Partial budgeting can also be used to estimate the cost benefits of introducing new feeding equipment or feeding methods. An analysis of the costs and benefits for this would involve a comparison of capital, operating, maintenance, and depreciation costs for the new equipment, associated changes in labor requirements, and a determination of any changes in growth rates and food conversion efficiency resulting from the proposed change in feeding method. As much information as possible should be gathered for use in partial budgeting. In addition to information from supply companies, details and experiences should be sought from other farmers, and from advisory and extension agencies wherever possible. It is very common to see redundant feeding systems on fish farms, the farmer having either changed system or reverted to hand feeding. Careful investigations should be made in an attempt to avoid such costly mistakes. Some suppliers are prepared to lend equipment to farmers for on-site trials, and comparisons with existing feeding equipment and methods.

Annual Profit Index

Where feeds are being tested within the limits of commercial production, it may be necessary to determine the performance of a new food type over a complete production cycle. An annual profit index, using gathered data, can be used to measure the performance of new feeds within the context of overall farm costs (New and Wijkstrom, 1990).

Unit feed costs and periodic FCRs may be inadequate as measures of the commercial value of a feed formulation. This is particularly the case in semi-intensive production systems, where the productivity of individual ponds is often variable, and where natural food organisms make an undetermined contribution to the feeding dynamics within the pond. Under these circumstances an annual profit index, based on the annual value of the pond (tank or cage) output, divided by the total cost of feeding, may be a better criterion by which to assess feed types. Annual profit indexes should take into account (From New and Wijkstrom, 1990):

A = Annual fish or shrimp yield (includes growth rate, number of harvests, down time, etc. (metric tonnes (MT)/ha/yr).

B = Annual quantity of feed used (MT, dry matter basis/production unit).

C = Unit feed cost (US$/MT dry matter).

D = Cost of all other inputs per unit of fish produced (because the type of formulation of the feed used may affect many other costs, labor, power costs, storage etc. (US$/MT).

E = Market value of the fish or shrimp produced (value, not weight important to farmer (US$/MT).

where:

$$\text{Annual Profit Index} = \text{API} = A \times E/(B \times C)/(A \times D)$$

COST-EFFECTIVE RATIONS

While a farmer can calculate the cost of growing fish from feed alone, based on food conversion values and feed costs, there are wider implications when feed costs are measured against farm output levels. Most farms have a fixed carrying capacity, limited by pond, tank, or cage volume, and water exchange rates. If the farmer, through skilled management, can produce more fish from these facilities, then the ratio of the fixed operating costs in relation to revenue will decrease.

The potential to produce more fish is linked to growth rate, which the farmer regulates by choice of feed type and ration size. In Chapter 8 it was seen that ration sizes can be selected for maximum food intake and growth, or set at restricted levels which may offer improved food conversion efficiency at the cost of some growth. Farmers must therefore choose ration sizes which meet production objectives, and yet remain cost effective. Feeding to satiation for maximum growth may optimize the use of facilities in terms of production, but requires careful control to avoid wastage. Conversely, feeding restricted rations may reduce wastage problems but at the loss of growth and production. Effective decisions concerning ration size should ideally be based on a knowledge of the growth–ration relationships for any particular operation (Bjørndal and Uhler, 1993). Where this information is not available, then FCRs are the farmer's best indicators of overall feeding performance, reflecting problems arising from either overfeeding or underfeeding. Improvements in feed quality and feeding practices have resulted in overall reductions in the FCRs obtained in many sectors of the industry. This is reflected in shorter production times and increased production of fish or shrimp per unit of feed used. All farmers should be aware of the potential FCRs attainable with particular feeds and species, and the values obtained within any operation should be taken as key indicators of the efficiency of feeding operations.

FOOD WASTAGE

Any reduction of food wastage will have a significant, beneficial impact on feed costs. Overfeeding is the most common source of wastage, while poor storage, high dust and fines levels in feeds, are also contributing factors. High growth rates accompanied by high FCR values are indicators of food wastage resulting from overfeeding. Under these circumstances the quality of the food, ration sizes, and feeding methods should be thoroughly examined.

A review of feeding practices should be conducted periodically on all farms, and staff should be kept fully updated on new developments. No farms can afford the losses generally associated throughout the industry with uneaten food. In those countries with highly competitive industries and strict environmental regulations, farmers and feed suppliers have significantly revised feed formulations and feeding practices. More improvements in waste control measures can be anticipated. Particularly useful will be the further development of practical equipment, capable of monitoring the biomass of fish and shrimp populations, and of detecting any wastage which occurs during feeding.

REFERENCES

BJØRNDAL, T. and R.S. UHLER. 1993. Salmon sea farm management: Basic economic concepts and applications. In Salmon Aquaculture, eds. K. Heen, R.L. Monahan, and F. Utter, pp. 239–254. Fishing News Books, Oxford.

CRAMPTON, V. and A. JACKSON, 1981. Don't lose out in the food chain. Fish Farmer 4(3):31–33.

EASLEY, J.E. 1977. Response of costs and returns to alternative feed prices and conversions in aquaculture systems. Marine Fisheries Review **39**:15–17.

JOLLY, C.M. and CLONTS, H.A., 1993. Economics of Aquaculture. Haworth Press, Inc., New York.

MAGUIRE, G.B. 1992. Issues associated with the evaluation of commercial prawn farming diets. In Proceedings of the Aquaculture Nutrition Workshop, Salamander Bay, April 15–17, 1991, eds. G.L. Allan and W. Dall, pp. 237-240. NSW Fisheries, Brackish Water Fish Culture Research Station, Salamander Bay, Australia.

NEW, M.B. and U.N. WIJKSTROM. 1990. Feed for thought. World Aquaculture **21**:17–23.

SHANG, Y.C., 1990. Aquaculture Economic Analysis: An Introduction. Advances in World Aquaculture, Vol. 2, Ed. P.A. Sandifer. World Aquaculture Society, Baton Rouge.

SMITH, P. 1990. Innovations in salmon and shrimp feed. Aquaculture International Congress Proceedings, pp. 121–126. Aquaculture International, Vancouver.

Index